AF688740

Einstein's Revolution: A Study of Theory Unification

Authored by

Rinat Nugayev
Volga Region State Academy, Kazan,
The Republic of Tatarstan,
The Russian Federation

BENTHAM SCIENCE PUBLISHERS LTD.

End User License Agreement (for non-institutional, personal use)

This is an agreement between you and Bentham Science Publishers Ltd. Please read this License Agreement carefully before using the ebook/echapter/ejournal (**"Work"**). Your use of the Work constitutes your agreement to the terms and conditions set forth in this License Agreement. If you do not agree to these terms and conditions then you should not use the Work.

Bentham Science Publishers agrees to grant you a non-exclusive, non-transferable limited license to use the Work subject to and in accordance with the following terms and conditions. This License Agreement is for non-library, personal use only. For a library / institutional / multi user license in respect of the Work, please contact: permission@benthamscience.org.

Usage Rules:

1. All rights reserved: The Work is the subject of copyright and Bentham Science Publishers either owns the Work (and the copyright in it) or is licensed to distribute the Work. You shall not copy, reproduce, modify, remove, delete, augment, add to, publish, transmit, sell, resell, create derivative works from, or in any way exploit the Work or make the Work available for others to do any of the same, in any form or by any means, in whole or in part, in each case without the prior written permission of Bentham Science Publishers, unless stated otherwise in this License Agreement.
2. You may download a copy of the Work on one occasion to one personal computer (including tablet, laptop, desktop, or other such devices). You may make one back-up copy of the Work to avoid losing it. The following DRM (Digital Rights Management) policy may also be applicable to the Work at Bentham Science Publishers' election, acting in its sole discretion:

- 25 'copy' commands can be executed every 7 days in respect of the Work. The text selected for copying cannot extend to more than a single page. Each time a text 'copy' command is executed, irrespective of whether the text selection is made from within one page or from separate pages, it will be considered as a separate / individual 'copy' command.
- 25 pages only from the Work can be printed every 7 days.

3. The unauthorised use or distribution of copyrighted or other proprietary content is illegal and could subject you to liability for substantial money damages. You will be liable for any damage resulting from your misuse of the Work or any violation of this License Agreement, including any infringement by you of copyrights or proprietary rights.

Disclaimer:

Bentham Science Publishers does not guarantee that the information in the Work is error-free, or warrant that it will meet your requirements or that access to the Work will be uninterrupted or error-free. The Work is provided "as is" without warranty of any kind, either express or implied or statutory, including, without limitation, implied warranties of merchantability and fitness for a particular purpose. The entire risk as to the results and performance of the Work is assumed by you. No responsibility is assumed by Bentham Science Publishers, its staff, editors and/or authors for any injury and/or damage to persons or property as a matter of products liability, negligence or otherwise, or from any use or operation of any methods, products instruction, advertisements or ideas contained in the Work.

Limitation of Liability:

In no event will Bentham Science Publishers, its staff, editors and/or authors, be liable for any damages, including, without limitation, special, incidental and/or consequential damages and/or damages for lost data and/or profits arising out of (whether directly or indirectly) the use or inability to use the Work. The entire liability of Bentham Science Publishers shall be limited to the amount actually paid by you for the Work.

General:

1. Any dispute or claim arising out of or in connection with this License Agreement or the Work (including non-contractual disputes or claims) will be governed by and construed in accordance with the laws of the U.A.E. as applied in the Emirate of Dubai. Each party agrees that the courts of the Emirate of Dubai shall have exclusive jurisdiction to settle any dispute or claim arising out of or in connection with this License Agreement or the Work (including non-contractual disputes or claims).
2. Your rights under this License Agreement will automatically terminate without notice and without the need for a court order if at any point you breach any terms of this License Agreement. In no event will any delay or failure by Bentham Science Publishers in enforcing your compliance with this License Agreement constitute a waiver of any of its rights.
3. You acknowledge that you have read this License Agreement, and agree to be bound by its terms and conditions. To the extent that any other terms and conditions presented on any website of Bentham Science Publishers conflict with, or are inconsistent with, the terms and conditions set out in this License Agreement, you acknowledge that the terms and conditions set out in this License Agreement shall prevail.

Bentham Science Publishers Ltd.
Executive Suite Y - 2
PO Box 7917, Saif Zone
Sharjah, U.A.E.
Email: subscriptions@benthamscience.org

CONTENTS

FOREWORD

Dr. Rinat M. Nugayev is a professor of department of humanities & social sciences at the Volga Region State Academy, Kazan 35, Universiade Village, 420138, the Republic of Tatarstan, Russian Federation. His book "*Einstein's Revolution: A Study of Theory Unification*" exhibits an attempt to provide a rational reconstruction of Einstein's Revolution that took place at the beginning of the XXth century.

In my view, his attempt can be taken as interesting and successful but not least because the reconstruction is grounded on his lucid epistemological mature theory-choice model. The model tries to advance the ideas of Russian physicists and philosophers of science M.I. Podgoretzky and J.A. Smorodinsky that hinge upon the "cross-contradiction" concept. I reckon that Nugayev's perspicuous account, embracing vast history of science stuff, is more convincing in many important respects than the corresponding eminent reconstructions of Thomas S. Kuhn, Imre Lakatos and their partisans.

The problem of scientific revolutions is still one of the most significant problems of the contemporary epistemology and philosophy of science, hence the author's book is up to date. Thus, I uphold the view that Rinat M. Nugayev's book should be published.

Elena Mamchur
Department of Philosophy of Science,
Institute of Philosophy, Russian Academy of Sciences,
12/1 Goncharnaya str., Moscow, 109420,
Russian Federation

PREFACE

It is a commonplace that Albert Einstein's scientific contributions were highly motivated by the Ideal of Unity of physical laws, and this had a considerable influence on the whole theoretical physics community. However, all the scientific career of Einstein after 1915, *i.e.* after the general theory of relativity had been achieved, was precisely the quest for unitary theories, unification of gravitation and electrodynamics, *etc.* And it is no wonder that the idea of unity of nature is best illustrated by *these* attempts of Einstein towards unitary theories during almost forty years than by the early works.

Yet, in my innermost conviction, Einstein's mature unification efforts and especially his general relativity are grounded on his early works and substantially on his 1905 strenuous efforts to create not only special relativity but the early quantum theory as well.Hence *the overall aim of the present book is to unfold Einstein's unificationist modus operandi, the hallmarks of actual Einstein's methodology of unification that engendered his 1905 special relativity and 1905 audacious hypothesis of light quanta.*

It turns out that in the particular case of Einstein, a discussion of methodology is unavoidable: Einstein expressed himself with regular reference to his "method", not unlike Poincaré, Planck and other physicists of his generation.

To achieve the object, a lucid and comprehensible epistemic model is exposed aimed at an analysis of the reasons for mature theory change in science (**chapter1**). According to the epistemic model, scientific revolutions or *radical breakthroughs in science were not due to ingenious invention of new paradigms or the fanciful creation of new ideas 'ex nihilo', but rather to the long-term processes of the reconciliation, interpenetration and intertwinement of 'old' research traditions preceding such breaks.*

Accordingly. origins of scientific revolutions lie not in a clash of fundamental theories with facts, but of "old" mature research traditions with each other, leading to contradictions that can only be attenuated and even eliminated in a more general theoretical approach. This is not to derogate the role of experiments in science. On the contrary, the proposed epistemic model seems to elaborate further the point of view stated in the current philosophy of science literature that both theorists and experimentalists have breaks in their respective traditions, but they are not typically simultaneous. Theory development must have, to some extent, a life of its own. The development of two main cultures within science does not mean that the two do not speak to each other.

The epistemic model proposed is illustrated and elucidated with reference to the Copernican Revolution (**chapter1**).

In **chapter 2** the model is implemented to amend the history of Special Theory of Relativity. It is contended that to figure out in what way different parts of Einstein's 1905 'annus mirabilis' writings hang together one has to take into account neo-Kantian epistemology. In this book the methodology is approached as an inalienable, active component of Einstein's engagement with his science.

The most influential Kantian concept necessary to conceive Einstein's Special Theory of Relativity creation and all his 1905 writings *as a whole*, as well as the *order* of their arrangement is Kant's *regulative* idea of the systematic Unity of Nature. It is maintained that Special Theory of Relativity turns out to be a mere stage of implementation of maxwellian electrodynamics, statistical mechanics and classical thermodynamics reconciliation programme. Since the ether conception was a substantial snag for Einstein's statistical thermodynamics design, the leading part in the breakthrough was played by Einstein's 1905 light quanta hypothesis.

Correspondingly, the arguments are exhibited in favor of the necessity to modify the history of the genesis and advancement of General Theory of Relativity. It is maintained that the dynamic creation of General Theory of Relativity had been incessantly governed by internal tensions between the two classical research traditions, that of special relativity and Newton's gravity (**chapter 3**).

The encounter of the traditions and their interpenetration and intertwinement entailed construction of the hybrid domain at first with an irregular set of theoretical models. Step by step, on eliminating the contradictions between the models contrived, the hybrid set was put into order. It is stressed that the main reason for the victory of General Theory of Relativity over the ingenious rival programmes of Abraham and Nordström was a synthetic character of Einstein's programme. Moreover, Einstein had put forward as a basic synthetic principle the principle of equivalence that radically differed from that of rival approaches by its open, flexible and contra-ontological character.

CONFLICT OF INTEREST

The author (editor) declares no conflict of interest, financial or otherwise.

ACKNOWLEDGEMENTS

It is a pleasure to thank Ms. Humaira Hashmi, Editorial Manager of Publications, and Ms. Hira Aftab of the eBooks Publication Department for their valuable assistance. My love and gratitude to my wife Gulnur and children Shamil and Nafisa for help and encouragement.

Rinat Nugayev
Volga Region State Academy, Kazan,
The Republic of Tatarstan,
The Russian Federation

CHAPTER 1

Mature Theory Unification and the Objectivity of Scientific Knowledge

Abstract: Multifarious arguments pro et contra convergent realism are scrutinized. It is exhibited that to overcome the troubles of convergent realism one has to turn from classical or 'metaphysical' realism to non-classical or 'internal' one, grounded on Kantian epistemology, and to the so-called 'coherent theory of truth'. Internal realism has no unassailable problems in grappling with the problem of empirically-equivalent theoretical descriptions and historical meta-induction problem, but obtains the quandary of objectivity of scientific knowledge instead. One of the ways out is proposed by the scientific knowledge growth model elaborated by Rinat Nugayev and by Peter Galison. Any paradigm is a local viewpoint that springs out of the culture into which its creator was submerged. However, an encounter of different paradigms leads to their interaction and intertwinement and, as a result; the crossbred theoretical objects are carefully constructed. *Via* their systems, the interpenetration of one paradigm into the other one takes place. To elucidate the proposed model of the growth of knowledge, the inter-theoretical context of the Copernican revolution is scrutinized.

Keywords: Almagest, Aristotle, Barnes, Bloor, Carnap, Cartwright, Coherence theory of truth, Convergent realism, Copernican revolution, Crossbred objects, Divine astronomy, Entity realism, Feyerabend, Foucault, Galileo, Galison, Hacking, Hybrid theory, Internal realism, Kantian epistemology, Kepler, Koertge, Kuhn, Lakatos, Laudan, Mature theory, Maxwell.

1. INTRODUCTION: CONVERGENT REALISM: PRO ET CONTRA

One should not be surprised that, after the untimely demise of logical positivism, scientific realism has returned as a pivotal philosophical stance. Yet some epistemologists and philosophers of science still dare to work within the tradition of Carnap, Hempel *et al.*, meticulously formalizing ideal patterns of scientific discourse. But the majority became gradually aware that this approach has vague relation to the reality of modern scientific practice, to the real problems the men of modern science are grappling with. Hence they turned instead, following Paul Feyerabend, Thomas Kuhn, Imre Lakatos *et al.*, to *historical* analysis for insight into the structure and dynamics of scientific theories. While contemporary thinkers stubbornly retain the logical empiricists' concern with issues of scientific

Rinat Nugayev
All rights reserved-© 2018 Bentham Science Publishers

objectivity, they are no longer adamantly committed to the logical empiricists' excessively formal account of what this objectivity consists in.

On the other hand, with hindsight, many thinkers, starting from James Stuart Mill and up to Rudolf Carnap, opted for a 'phenomenal' account of the content of scientific propositions. They boldly contended that terms like 'mass', 'charge' and 'force' were properly comprehended as standing for 'complexes of observable circumstances'. Nevertheless, the rise of modern quantum theory has made this entrenched stance especially problematic. In particular, the strenuous attempts to reduce such theoretical discourse to observational claims ran into blatant technical obstacles [1]. Indeed, it is one thing to maintain that claims about tables and chairs (or rods and clocks) are really claims about sensations. But it is hard to take this tenet seriously for assertions about unobservable objects like isospins, charges and colors of up and down quarks, quantum gravity loops, superstrings, and so forth.

It is a commonplace that realism is rooted in the conjunction of the two basic tenets.

I. *An independence thesis:* Our judgements answer for their truth to a world which exists independently of our awareness of it.
II. *A knowledge thesis:* We can know which of these judgements are true.

Alas, the history of philosophy informs us that a classical counter-argument against realism was worked out due to Rene Descartes: if the external world is independent of our knowledge of it, how trustworthy knowledge about it is possible? – And all the further problems with realism hinge upon this counter-argument to a certain extent.

Now let us turn to the so-called '*scientific realism*' that had showed up within the realist epistemological research programme with the "hard core" moulded by the (I) – (II) tenets. Scientific realism is prompted by the assertion that one should believe in the existence of the objects postulated by our most *successful* scientific theories like Newtonian mechanics, maxwellian electrodynamics, relativity or quantum theory. Science is the highest authority to guarantee the existence not only of an "electron' but of a 'black hole", "naked singularity", "M-brain" and the "Multiverse" as well.

It is proverbial that the most convincing argument in defense of "scientific realism" is the "*no miracles*" one. Just as Hilary Putnam maintained, "the positive argument for realism is that it is the only philosophy that doesn't make the success of science a miracle". However, the 'no miracle' argument is as a matter of fact an '*inference to the best explanation*' or Pierce's '*abduction*'. But, as Larry

Laudan contended, since the anti-realists – like Bas van Fraassen ([2]) *et al.* – stubbornly reject the 'inference to the best explanation' in natural science, it is quite improbable that they can bear it on the meta-level.

Nevertheless, the two following strong counter- arguments against scientific realism have been elaborated.

1.1. The Tenet of Underdetermination of Theory by Observational Evidence

Suppose that two theories T_1 and T_2 (*e.g.* Ptolemaic and Copernican astronomies, or special theory of relativity and Lorentz's electron theory) are empirically-equivalent, in the sense that they make the same observational predictions. Then no body of observational evidence (*e.g.* the notorious Michelson-Morley experiment or anomalous Mercury perihelion shift) will be able to decide conclusively between T_1 and T_2.

The proverbial answer, according to which we should always choose the theory that provides the better explanation, unavoidably plunges into historical obstacles. If we delve into the cases where natural scientists have embraced their 'best explanations' as true, these explanations have commonly turned out false. Consider, for instance, Ptolemaic astronomy, the caloric theory of heat, the ether theory in electrodynamics of moving bodies, and so on.

1.2. The Tenet of Pessimistic Meta-induction from Past Falsity

This sober anti-realist argument is grounded on the analysis of mature theory change in science (the so-called 'scientific revolutions'). According to the meta-induction argument, tentative reflection on the abandonment of theories in the history of science provokes the expectation that our best current scientific theories will themselves be inexorably abandoned.

In his thought-provoking paper "*A Confutation of Convergent Realism*" Larry Laudan [3] famously unfolded the basic premises of the so-called 'convergent realism'.

(R_1) Scientific theories are typically *approximately true*, and more recent theories are closer to the truth than the older ones in the same domain (*e.g.* Newtonian mechanics and relativistic mechanics).

(R_2) The observational and theoretical terms within the mature scientific theories genuinely *refer* (an 'electron', a 'black hole', a 'gravitational wave', a 'naked singularity', *etc.*).

(R_3) Successive theories in any mature science will be such that they preserve the theoretical relations and the referents of earlier theories so that earlier theories will be limiting cases of later ones (classical mechanics and quantum mechanics, classical electrodynamics and quantum electrodynamics, general relativity and special relativity).

(R_4) Acceptable new theories should perspicuously explain why their predecessors were successful in so far as they were successful (*e.g.* the so-called 'weak-field approximation in general relativity').

(R_5) Tenets (R_1) to (R_4) entail that scientific theories should be successful.

Note that a theory is considered 'successful'

a. if it makes substantially correct predictions (Mercury perihelion shift);
b. if it leads to efficacious interventions in the natural order (transistors, A-bombs, lasers);
c. if it successfully passes a host of standard tests (three critical tests of general relativity).

As a result, the aforementioned tenets represent one of the best explanations of the successes of science.

However, a more thorough analysis (partly grounded on Laudan's advances) enables one to take the (R_1) to (R_4) tenets with a grain of sault.

A. As Thomas S. Kuhn had famously exhibited [4 - 6], tentative comparison of the *Weltanschauungen* of Aristotle, Newton, Einstein, Bohr and Witten enables one to arrive at a conclusion that they are not like the snapshots of one and the same object made with the different orders of exposure. In what sense Aristotle's physics is an approximation of Newton's physics? And, likewise, is Newton's physics an approximation of Einstein's physics?
B. Let us delve into the state of ethereal theories in the 1830s and 1840s. It was commonly believed that the electrical fluid constitutes a substance able to accumulate on the surface rather than to permeate the interstices of bodies. This concept had been applied to conceive the attraction of oppositely charged bodies, the queer behavior of the Leyden jar, the apparent similarities between atmospheric and static electricities, and so on. Within chemistry and heat theory, the caloric ether had been widely accepted and used since H. Boerhave (by Lavoisier, Laplace, count Rumford *et al.*) to comprehend everything from the role of heat in chemical reactions to the conduction and radiation of heat. Within an undulatory theory of light, the optical ether concept was an apparent success in comprehending reflection, refraction, interference, diffraction and

polarization, not to forget that the optical ether theories had famously predicted quite new phenomena. The most blatant example is A. Fresnel's prediction of a bright spot at the center of the shadow of a circular disc [7]. Hence it is worth recalling James Clerk Maxwell who famously claimed that the ether was better confirmed "than any other theoretical entity in natural philosophy".

C. In science there were (the ether theory) and there are (the theory of superstrings) empirically and theoretically successful theories containing non-referential terms. For instance, according to Dawid Richard [8] "The theoretical posits of string theory get their meaning only relative to their theoretical framework and must be understood as mathematical concepts without any claim to 'corporal' existence in an external world. The world of superstrings cut all ties with classical theories about physical bodies [...] Neither the established brands of scientific anti-realism nor the conventional ontological formulation of scientific realism are compatible with spirit and content of String Physics".
D. Some eminent 'convergent realists' (Heinz Post or Noretta Koertge, for instance) maintain that 'new' theories retain the confirmed portions of the 'old' ones. But the Copernican astronomy did not retain the key mechanisms of the Ptolemaic one like the notorious 'equant'. Newton's physics did not retain even most of the theoretical laws of Cartesian mechanics, astronomy and optics. Franklin's electrical theory did not contain its predecessor - Nollet's subtle theory – as a limiting case. Einstein's special relativity and his theory of quanta did not retain the luminiferous ether and especially the radiation mechanisms associated with it. Moreover, the general relativity was in some sense a return to the ether (Einstein's correspondence with H.A. Lorentz).
E. Whatever the intentions of convergent realists, their assertions that after a scientific revolution the laws of the 'old' theory magically transform into the limiting cases of the 'new' theory reflect their best wishes only. Of course it remains unclear *why* on earth new theoretical ontology should smoothly transform into the old one when some peculiar parameters change; for instance, under $(v/c) \rightarrow 0$ or $(S/h) \rightarrow \infty$ (v- velocity of the body, c – velocity of light, S – action, h – Planck's constant).

Note that the superposition principle in quantum mechanics does not transform into classical one under $(S/h) \rightarrow \infty$, and 'mass' does loose its ability to transform into 'energy' under $(v/c) \rightarrow 0$. Likewise, consider the transition from Einstein's nonlinear field equations in general theory of relativity to the so-called 'weak-field approximation': it is insufficient to leave the first members of the metric expansion. One has to introduce some *substantial* assumptions that transform the new ontology into the 'old' one [9].

Of course this does not exclude the peculiar cases when *some* consequences of the new theory transform into the consequences of the old one, as is the case for classical and relativistic mechanics. For instance, the length of a moving body $l = l_0 \sqrt{1 - (v/c)^2}$ under $(v/c) \rightarrow 0$ transforms into l_0.

As Paul Feyerabend had put it in "*Against Method*", even Niels Bohr himself, being the author of the 'correspondence principle'(1922), had warned that the 'asymptotic connection' between the classical physics and the quantum one,

> "As it is interpreted in the correspondence principle, does not lead to the gradual elimination of the differences between quantum understanding of the radiation phenomena and the ideas of classical electrodynamics; all that is asserted is merely an asymptotic reconciliation of numerical statistical results".

As a result, according to Larry Laudan, in spite of the fact that convergent realism was conjured up to comprehend the successes of science, his results constitute as a matter of fact hypotheses ad hoc on the meta-level:

> "There is a [lucid] difference between wanting to believe something and having good reasons for believing it" ([3], p. 137).

On my view, convergent realism is most likely a myth, that constitutes the core of 'scientific ideology' (see for instance [10]) concocted to disguise the contradictions between the main '*Weltanschauungen*' of modern science. Yet it can be helpful in bringing up the new generations of young *Naturforschers*.

In my judgement *the Convergent Realism should be substantially modified* at the expense of rejecting one or several assertions $(R_1) - (R_5)$ of the protecting belt of the Realist Epistemological Research Programme and *via* introduction of some new bold hypotheses.

2. CONSPICUOUS ALTERNATIVES OF CONVERGENT REALISM

(2.1) Nancy Cartwright in "*How the Laws of Physics Lie*" [11] and Ian Hacking in "*Representing and Intervening*" [12] have worked out an eminent standpoint that they call 'entity realism', as opposed to 'theory realism'. Their approach maintains that the modern physics is responsible for any number of striking physical effects, from lasers to electron microscopes, superconductors and supercomputers. However, it resolutely denies that these physical effects provide convincing support for modern mature physical theories. In particular, Nancy Cartwright punctuates that the standard 'derivations' of the above mentioned effects from 'fundamental' theory are necessarily mediated by dubious ad hoc assumptions, mathematical short cuts, and 'fudge factors'. Since these subtle and deft devices are in general unmotivated by the fundamental theory, the derivation fails to provide any inductive support for the fundamental theory. Thus, lasers and superconductors are as a matter of fact Kantian 'noumena' ('things in themselves'), and quantum electrodynamics has nothing to do with them.

Yet this does not imply the non-existence of the subatomic particles and other unobservable Ω-hyperons implicated in the relevant physical effects. It means that the success of our deliberate attempts to manipulate these entities is ample testimony to their existence. As Ian Hacking famously contended, "If you can spray them, they are real". In my view, N. Cartwright and I. Hacking have pioneered the epistemological problem of relevance of the applied sciences to the problem of truth. Yet their final conclusion building the 'iron curtain' between the fundamental, partial and empirical laws is rather dubious. Certainly artful simplifications play an important role in science. But many of them are themselves guided by theoretical considerations, and so add to the credit of the theory when they succeed ([1], 19). For instance, in a 1974 lecture Werner Heisenberg famously recalled that Einstein had confessed to him that "Perhaps I did use such philosophy [Machism] earlier, and also wrote it, but it is nonsense all the same...*it is in fact the theory which first determines what can be observed"* (quoted from [13],169; my italics).

Any scientist who used the approximation methods in his research practice is aware of the following. To get an approximate solution in the case when the basic equation cannot be solved, one has to expand the solution in a power series of some parameter (for instance in v/c in relativistic mechanics). However, one has to determine *beforehand* what component of the expansion should be left. But it is theory itself that decides.

For instance, in general relativity the general solution of Einstein's 16 intricate field equations is lacking. But one can obtain an approximate solution in the so-called 'weak-field approximation' for deriving some consequences of Newton's theory of gravitation. To arrive at an approximate solution one has to expand the metric in a power series of gravitational potential and then neglect all the square-law components and the components with higher degrees. But how can one arrive at an approximate solution without Einstein's equation themselves? Or consider the famous spherically-symmetric Schwarzchild solution of Einstein's equations that was derived by Karl Schwarzchild. The solution was obtained in the trenches of the world war only in 1916 – only *after* Einstein had posited his famous equation at the Prussian Academy of Science in November 1915.

(2.2) John Worrall holds a different stand to 'save' realism that looks more fruitful and promising. He concedes that even predictively successful past theories often contain significant errors. Nevertheless, he maintains that this does not require a blanket rejection of all scientific claims about the unobservable mechanisms behind the observable predictions. He asserts that history of science exhibits that past theories are characteristically wrong about the *nature* of the unobservable realm, but not about the *structure* of its behavior. In particular in his pivotal example he argues that XIX-th century scientists were blatantly wrong to think of electromagnetic radiation as embodied in a luminiferous 'ether', but quite right about the mathematical equations governing electromagnetism. As a result we should believe in the *structure* of unobservable reality postulated by successful theories, but avoid committing ourselves to any claims about the *nature* of that reality.

Just as Poincaré put it in his epoch-making "*Science and Hypothesis*",

> "for Fresnel's object was not to know whether there really is an ether, if it is or is not formed of atoms, if these atoms really move in this way or that; the differential equations are always true... these equations express relations, and if the equations remain true, it is because the relations preserve their reality" (quoted from [14], p. 157).

From the vantage point of view of subsequent theories Fresnel completely misapprehended the "nature of light"; but, nonetheless, it is no wonder that his theory enjoyed the empirical predictive success that it did. It is 'no miracle' because Fresnel's theory, as science later was convinced, attributed to light the right structure.

According to Bas van Fraassen, it is the growing mathematization of science that paves the way for structuralism. Its motto remain Poincaré's words:

> "The aim of science is not things themselves, as the dogmatists in their simplicity imagine, but the *relations between things*; outside those relations there is no reality knowable".

Poincaré shrewdly took the unobservable entities postulated by scientific theories as Kant's 'things in themselves' [15]. Yet he was a blatant 'revisionist', arguing that the latter can be known *indirectly* rather than not at all since it is possible to comprehend the relations into which they enter. As a matter of fact Poincaré followed Edmund Husserl's brand of neo-Kantianism, beginning with the goal of recovering the objective or *intersubjective* world from the subjective world of private sense impressions:

> "what we call objective reality is... what is common to many thinking beings and could be common to all; ...the harmony of mathematical laws" ([16],p.14).

Note that a compelling hallmark of structural realism consists in that it follows the Galilean trend of theoretical physics. Galileo not only followed Plato's "*Timeus*", contending that "the Book of Nature is written in mathematical language". In a clash with Aristotle he tried to lead physics away from grasping the 'essences of things'. As Galileo had famously put it, "the search for essences, in my judgement, is a vain and hopeless kind of pursuit".

On the other hand, there are plenty examples of continuity in the mathematical structure of successive scientific theories. For instance, Niels Bohr and others famously applied the "correspondence principle" disclosing that quantum-mechanical mathematical models ought to reduce to the classical ones in the limit of large numbers of particles. Furthermore, there are numerable cases in quantum mechanics where the Hamiltonian functions that represent the total energy of mechanical systems imitate those of classical mechanics but with variables like those that stand for position and momentum replaced by Hermitian operator. In this connection Simon Saunders [17] unfolded how much structure even Ptolemaic and Copernican astronomies have in common.

Many partisans of structural realism contend that the most fruitful trends of its advancement are connected with gauge quantum field theories, which are our best contemporary physical theories of all the forces other than gravity. Each theory is associated with a different symmetry group, and the unification of theories was successfully achieved by looking for theoretical structures with combined symmetry. For instance, the unitary group U(1) for quantum electrodynamics, the special unitary group SU(2) for the theory of weak interactions, and SU(2) x U(1) $/Z^2$ for the unified electroweak theory were constructed according to the above mentioned pattern.

The fundamental objects of these theories are ontologically only secondary to structure since the objects of a theory are members of equivalence classes under symmetry transformations and no further indication of the objects is possible. This is especially plausible when it comes to quantum field theory in curved spacetimes because in that context "a useful particle interpretation of states does not exist" [18]. The notion of a particle turns out to be reference dependent. As in typical crossbred theory, there is a fundamental collision between relativistic quantum field theory and the existence of a localizable particle common for general theory of relativity.

One of the most blatant fields of applications of the structural realism appears to be modern theory of superstrings. Although there still doesn't exist any direct experimental evidence for superstrings, it is acknowledged by a majority of high energy physicists as the only promising candidate for the construction of the "Theory of Everything", of a unified theory of all the four basic forces. However, superstring theory's lack of experimental confirmation and its incomplete state are generally acknowledged as deplorable weak points. Nevertheless, a theory may be corroborated on purely theoretical grounds. It may provide subtle advances of the theoretical comprehension of important interconnections between already known physical phenomena. Accordingly, string theorists provide a number of amazing instances of conceptual coherence. Three of the most startling ones are:

1. the fact that a theory of quantized extended object implies gravity;
2. the fact that a string theory of matter necessarily requires supersymmetry;
3. the fact that string theory in certain special cases can comprehend the connection between black hole entropy and the black hole event horizon.

Withal, the theory of superstrings is of especial need of subtle "scientific ideology" capable of uniting all the specialists and defending them from vitriolic attacks of their opponents. Hence

> "Neither the established brands of scientific anti-realism nor the conventional ontological formulation of scientific realism are compatible with spirit and content of String Physics [...] Since a theory's closeness to the fully universal and consistent theoretical schemes, consistent structure realism can provide a meaningful notion of realism without relying on claims about the truth or approximate truth of current scientific theories" ([8], p. 36).

Nevertheless, some substantial objections to the Structural Realism were put forward.

1. According to Psillos and Papineau [19], restriction of belief to structural claims is in fact no restriction at all.
   Indeed, a structure [20] S consists of:
   i. a non-empty set U of individuals (or objects) U, which constitute the domain of the structure, and
   ii. a non-empty indexed set R (*i.e.* an ordered list) of relations on U.

For instance, the electric field at a point **E**, the magnetic field at a point **H**, and the current density **J** are the basic theoretical objects of Maxwellian electrodynamics. Maxwell's equations exhibit the relations between them.

The relations between the theoretical objects do not exist without the objects themselves; hence propositions about the objects are at the same time the propositions about the structures, and *vice versa*.

In general the basic model (or the Fundamental Theoretical Scheme) is the crux of any mature theory [21]. The Fundamental Theoretical Scheme (FTS) is a system of basic objects + the fundamental equations of the theory. Yet, to form a full-blood mature theory, one needs the partial theoretical schemes also, as well as the empirical schemes. And all of them are constructed from the FTS according to certain rules. In addition, one needs the operational procedures that provide the empirical justification of the objects, and so on. How are these rules transformed under the theory change? Are they conserved in theory change?

Thus the notion of structure is inseparably linked with a host of other notions. And one has to unfold the whole chain of mature theory functioning: structure + corresponding mechanisms.

2. Mathematical structures are often lost in mature theory change
   The most apparent example is, of course, the Copernican Revolution. The latter consisted not in Ptolemy-Copernicus transition only, but in the construction of the whole system of heliocentric astronomy grounded on the Newtonian mechanics. The Copernican research programme advancement was accompanied by gradual elimination not only of an 'equant', but of the whole subtle structure of epicycles and deferents that was crowned by Kepler's and Newton's laws.
   The second apparent example is the ether mechanical models with their subtle parts transposed from the hydrodynamics, *etc*. The further examples embrace de Sitter's universe models, rejected by Hubble's discoveries, the unified theories of Einstein, Weyl *et al.* with a valuable exception of the Kaluza-Klein theory, and so on.
3. Structural realism is too metaphysically revisionary.
4. Structural realism can't account for causation.

In my view, the critical arguments (1) – (4) cannot ultimately refute the hard core of research programme of structural realism. They appeal for its modification, though. *One of the possible ways out lies in artful fusion of the structural realism programme with other congenial approaches, first and foremost with the approaches that flourished within the neo-Kantian philosophical tradition of the XX-th century.* Indeed, the ultimate stumbling block of current philosophy of science is "to provide a theory of objectivity without a representational theory of truth" (Ian Hacking).

3. INTERNAL REALISM IN THE CONTEXT OF THE COHERENT CONCEPTION OF TRUTH

Since realist epistemological research programme was devised to explain the success of science, it remains substantially ad hoc with respect to that success. In my view a way out is to modify radically the hard core of the programme. In that respect one has to take into account Hilary Putnam's [22] encouraging writings referring to the so-called '*internal realism*'. Putnam's study starts with outlining the two philosophical perspectives (or "philosophical temperaments") in history of Western philosophy.

(I) Metaphysical Realism: The world consists of some fixed multitude of mind-independent objects. Moreover, there exists only *one* true and complete description of "the way the world is". Hilary Putnam aptly denotes this perspective as

> "the *externalist* perspective, because its favorite point of view is God's Eye point of view" ([22], p. 251; my italics).

(II) Internal Realism: The second perspective has no generally adopted label, hence Putnam refers to it as the *internalist* perspective. It is characteristic of this view to hold "what objects does the world consist of?" "It is a question that it only makes sense to ask within a theory or description. "The internalist philosophers contend that there is more than one "true" theory or description of the world." 'Truth', in an internalist approach, is a kind of idealized rational acceptability – some sort of ideal *coherence* of our beliefs with each other and with other experiences as those experiences are themselves represented in our belief system. Hence truth is not correspondence with mind-independent or discourse-independent state of affairs. There is no 'God's Eye point of view' that one can stubbornly access. There are only different (often antagonistic) points of view of actual persons reflecting various (often hostile) interests and purposes.

Internal Realism, that holds together subjective and objective aspects of truth, is evidently grounded on Immanuel Kant's epistemology. Reason is not an exact snapshot of Nature. On the opposite, the crux of Putnam's doctrine is continuous ***interaction*** of Reason and Nature.

> "Saying that something is red, or warm, or furry, is saying that it is so-and-so *in relation* to us, not how it is from a God's Eye point of view" ([22],p.60; my italics).

Internalism does not renounce the standpoint that there are experiential inputs to knowledge. Knowledge is not a bulk of information with no constraints except internal coherence. Yet it does deny that there are inputs which are not themselves to some extent shaped by our concepts, by the subtle vocabulary we apply to report and describe them, or any inputs which admit of only one single description, independent of all conceptual choices.

What makes a statement (or a set of statements) rationally acceptable is its coherence and fit. Coherence of 'theoretical' (or less experiential) beliefs with one another and with more experiential beliefs, not to forget coherence of experiential beliefs with theoretical beliefs.

Withal, truth is not equal to rational acceptability for the following reason. Truth is supposed to be a property of statement that cannot be lost, whereas justification can. In a sense truth constitutes a certain limit, an idealization of rational acceptability. One speaks as if there were such contraptions as epistemically ideal conditions, and one calls a statement 'true' if it would be justified under such conditions.

The second apparent hallmark of the truth concept of internal realism is its **coherent** character. The multitude of its adherents embraces not only the thorough neo-Hegelians like William Bradley, Henri Bosanquet and Brand Blanschard, but such eminent analytical philosophers as Otto Neurath, Carl Hempel, Willard Quine, Donald Davidson and of course Hilary Putnam himself. For instance, according to H.H. Joachim, truth is "the *systematic coherence* which characterizes a significant whole" (quoted from [23], p.99; my italics). As a matter of fact, truth *is* the coherence, or mutual support, of our beliefs.

Note that, according to Ralph Walker [24], the coherent theorist does not depart so far from common sense as to have to deny such commonplaces as "true propositions correspond with the facts". The partisans of the coherent theory of truth can utter such remarks quite freely. Nevertheless, they just will not take them as expressing the *nature* of truth, nor will they consider the "facts" to belong to a metaphysically independent reality. On the contrary, on their view the facts are themselves determined by the coherent system of beliefs.

Moreover, to assert that the belief is true when it is a member of a coherent system is not to commit oneself to fanciful idealism. For the various elements of the relevant system are not restricted to beliefs only but also include non-subjective aspects of lived reality, such as common institutional practices, entrenched cultural traditions, and even the components of Husserl's 'lebenswelt' embracing the events of our everyday life.

A common objection to a coherent theory was famously fixed by Bertrand Russell ('bishop Stubbs' argument). According to Russell, a coherence theory allows *any* proposition to be true, since any proposition can be a member of some coherent set. Yet, according to Walker, Russell's objection misses the point, since coherence theorists maintain that truth consists in the coherence of some *specific* set of beliefs. Just as Charles Sanders Peirce has put it:

> "All the followers of science are fully persuaded that the process of investigation, if only pushed far enough, will give one certain solution to every question to which they can be applied. One man may investigate the velocity of light by studying the transits of Venus and the aberration of the stars; another by the oppositions of Mars and the eclipses of Jupiter's satellites; [...] They may at first obtain different results, but, as each perfects his method and his processes, *the results will move steadily together toward a destined center*. So with all scientific research. Different minds may set out with the most antagonistic views, but the progress of investigation carries them by a force outside of themselves to one and the same conclusion. This activity of thought by which we are carried, not where we wish, but to a foreordained goal, is like the operation of destiny. No modification of the point of view taken, no selection of other facts for study, no natural bent of mind even, can enable a man to escape the predestinate opinion. This great law is embodied in the conception of truth and reality. The opinion which is fated to be ultimately agreed to by all who investigate, is what we mean by *the truth*, and the object represented in this opinion is the real" ([25], 206; my italics).

'Coherentism' is a blatant alternative to *foundationalism*. The latter takes justification as having a structure like that of a tower, with certain beliefs serving as firm foundations and all others supported by them. On the contrary, 'coherentism' renounces this obsolete image and posits justification as having the structure of a raft. Justified beliefs, like the planks that successfully make up a robust raft, mutually support one another.

The other blatant hallmark of coherence theory of truth consists in that it rejects the Cartesian clear-cut distinctions between subject S and object O. Thus, according to Brand Blanschard ([26], 104),

> " If thought and things are conceived as related only *externally*, then knowledge is luck"(my italics).

This desire to transcend the common binary division between O and S, between the world, on the one hand, and human knowledge, experience, interpretive schemes, and the like, on the other hand, ascends to Kant and Hegel. In the Kantian and Hegelian (not to forget Heidegger) epistemological tradition, the first and most obvious problem with this binary picture is maintained to be that it is impossible for human beings to remove all traces of our engaged concern with the world. It is impossible for us mortals to attain a God's eye view. Hegel denounced this conclusion because he believed that the binary picture implicated in the peephole view is just as wrong in its metaphysics as in its epistemology.

> "We are not peering through a peephole at the world but are always already in its midst… and knowledge is the outcome of sensuous human practices involving some aspect of this world within which we live….for coherentism, knowledge is ultimately a product of phenomena that are *immanent* to human belief systems and practices, social organizations, and lived reality" ([27], 162; my italics).

This significant aspect of the coherent theory of truth was stressed by Michel Foucault who contended that truth isn't outside power, or lacking in power. "Truth is a thing of *this* world". Each society has a specific '*regime of truth'* of its own, its own 'general politics' of truth: that is, the types of discourse which it accepts and makes function as true.

Likewise, "by truth I do not mean the 'ensemble of truths which are to be *discovered* and accepted', but rather 'the *ensemble of rules* according to which the true and the false are separated and the specific effects of power attached to the true'" (Foucault [28], 318; my italics).

According to this standpoint, truth is being understood as not a matter of a battle 'on behalf of truth', but of a battle about the current *status* of truth and the economic and political role it plays.

For Foucault 'truth' is to be comprehended as a system of subtle procedures for the production, regulation, distribution, circulation and operation of statements. Hence 'truth' is linked in a circular relation with systems of power which produce and sustain it, and to effects of power which it induces and which extend it. However, contrary to Karl Marx and his followers, Foucault asserts that a 'regime of truth' is not merely ideological or 'super-structural'. Though, it was a necessary condition of the genesis and development of modern capitalism. As a result, it is not a matter of emancipating truth from every system of power (which would be a castle in the air), for "*truth is already power*".

What Foucault tried to exhibit is that the birth of clinic, penitential system and other notorious modern institutions was first and foremost the reflection of the growth of certain powerful ideologies. These ideologies, in turn, had interacted with wider ideological changes, especially with the demise of individualism in the XVII century, and so on. And his proposal is that the "clinic", the "asylum", the "prison", *etc.* are not proper ways of treating most folks and that our belief that they are a kind of ideological prejudice, or 'folly'. All in all, there is no *objective* place to stand in ideological battles.

However, in my judgement, one should not conclude from Foucault's startling account (elucidating the subtle dialectic of inner transformations of knowledge and power on micro-and-macro-levels) that scientific truth and ideology are identical. ***The ultimate aim of science is to produce Objective Knowledge.*** The growth of objective knowledge in science eventually consists in gradual 'purification' of knowledge from anthropic sociocultural, 'ideological' remnants. As Louis Althusser used to say, 'ideologies may be just or unjust, but not true or false'.

According to Foucault, history of mankind in general and history of science in particular constitute a series of 'discourses' or 'ideologies' that are changed due to irrational reasons. This point of view may be attractive for common history; yet for the history of science it seems dubious.

How on earth our most general beliefs arise? – Many modern thinkers maintain that the most compelling answers are provided by Marx, Freud and Nietzche. They take our religious and ethical ideas as grounded in the irrational, of mean class interest (Marx), of the underground unconscious (Freud), of the omnipresent 'will to power' (Nietzche).

Putnam stresses that such terms for rational acceptability as 'justified', 'well confirmed', 'best of the available explanations' and so on may be entirely subjective. However, the standpoint that rational acceptability itself is subjective is evidently a self-refuting one.

Although Putnam renounces 'unbridled relativism', coherent theory of truth necessarily contains a modicum of relativism. For instance, since the 'coherentness' of a view can be attained in many different ways. And this hallmark is tightly connected with another stumbling block: where the alternative ontologies (on which the empirically-equivalent theories are grounded) come from? According to the coherent account, they come into existence since we are engaged with and related to the external world in a variety of specific locations and with a variety of specific projects. But all these should not lead one to 'absolute relativism'.

> "Absolute relativism produces a tolerance for other cultures, but it also removes the need *to revise one's own beliefs and practices* in light of their difference from others" ([27], 175; my italics).

The actual situation one increasingly finds himself in rarely allows for a noncommittal tolerance. In global situation modern societies are so *interwoven* that few if any exist separate from the others; we are increasingly interdependent and *mutually influencing*.

For instance, the alternative ontologies of Ptolemy (mathematical astronomy) and Aristotle (qualitative physics) grew out of different (if not to say inimical) social practices. Ptolemy's ontology was grounded in pagan religious meditative practice (the divine mathematics of the Skies) and Aristotle's world view was grounded in sensual and object-manipulative one (the mundane physics of the Earth). For many centuries the dualism was 'frozen up' in the Aristotle-Ptolemy cosmology until Copernicus had blown the cosmology up from within. All these were performed in quite different (monotheistic) socio-cultural context; as a result Copernicus set up a host of various research practices aimed at the *reconciliation* of the physics of the Earth with the mathematics of the Skies.

4. INTERNAL REALISM AND THE OBJECTIVITY OF SCIENTIFIC KNOWLEDGE

Internal realism easily and cheerily solves the paradox of many empirically-equivalent descriptions of reality by punctuating that they come into existence since we are engaged with and related to the external world in a variety of specific locations and with a variety of specific practices. Consider for example the case of the Ptolemy-Aristotle research programme in astronomy.

This programme in astronomy was grounded on the motley pagan beliefs concerning the connections between a man and the world. They take place due to the activity of different capricious and revengeful gods that constantly quarrel with each other. The Man was treated by Greeks as a part of Nature, and the main problem of Ancient Science was to "save the phenomena" (Ptolemy), to seek for different and often inconsistent ways of fitting theory to observation data.

On the contrary, the Copernicus-Galileo-Newton research programme was grounded on the principles of monotheistic (Christian) theology that took Man and Nature as ingenious realizations of God's Design. The pivotal problem of Christian science (and -with certain reservations - of Moslem one) consisted in apprehending the Design as a whole, of comprehending the great Book of Nature written by Him in elevated mathematical language.

With similar degree of easiness internal realism solves the Second Paradox of historical meta-induction pointing out that the scientists belonging to changing paradigms "leave in different worlds" (T.S. Kuhn's expression) that are blatantly 'incommensurable' (Kuhn, Feyerabend). It is a platitude that the sequence of scientific internal '*Weltanschauungen*' should not tend to a certain limit.

However, as it often happens, on solving one set of problems one gets a pile of another conundrums. For instance, how the scientific progress should be understood? And what about the *objectivity of scientific knowledge*? Wherefrom should it take place within the internal realism and how can it grow?

In my view, the above mentioned problems can find judicious answers within a lucid epistemological model of theory change that I try to amend [29].

The model is grounded on the following natural assumptions.

1. One never comprehends the mechanism of mature theory change without grasping the mechanisms of theory unification that are tightly interconnected (see also [30]).
2. And we cannot apprehend the mechanisms of theory unification without the analysis of their genesis.

In certain respects the epistemological model leads us out of the strict borders of internal realism. Any paradigm, research programme, research tradition, *etc.* constituted a 'local', internal world view engendered by the peculiarities of the culture in which its adherent is immersed (Aristotle, Newton, Lorentz, Maxwell,

Einstein, Bohr *et al.*). Yet it is an *encounter* of different paradigms that leads to their real *interaction*, interpenetration and intertwinement. The contrivance of the crossbred theoretical objects, constructed from the basic objects of all the encountered theories, constitutes a necessary link of this interaction. The crossbred objects are intermediaries through which the interpenetration of the methods and notion of one paradigm into another takes place. As a result new theories emerge that reconcile with one another quite better.

It is no wonder that the make-up and the milestones of such a fusion were already scrutinized in Maxwell's seminal paper "*Hermann Ludwig Ferdinand Helmholtz*" that starts as follows:

> "Hence the ordinary growth of human knowledge is by accumulation round a number of distinct centers. The time, however, must sooner or later arrive when two or more departments of knowledge can no longer remain independent of each other, but must be *fused into a consistent whole*. But though men of science may be profoundly convinced of the necessity of such a fusion, the operation itself is a most arduous one. For though the phenomena of nature are all consistent with each other, we have to deal not only with these, but with the hypotheses which have been invented to systematize them; and it by no means follows that because one set of observers have labored with all sincerity to reduce to order one group of phenomena, the hypotheses which they have formed will be consistent with those by which a second set of observers have explained a different set of phenomena. Each science may appear tolerably consistent within itself, but before they can be combined into one, each must be stripped of the daubing of untempered mortar by which its parts have been prematurely made to cohere. Hence the operation of fusing two sciences into one generally involves much criticism of established methods, and the explosion of many pieces of fancied knowledge which may have been long held in scientific reputation" (Maxwell [31], 592; my italics).

It should be stressed that the process of reconciliation leads to the gradual attenuation or even elimination of all the anthropic, local components of scientific knowledge containing the remnants of contingent peculiarities of the genesis of the paradigm. (The most apparent example is the evolution of the concept of "force" beginning from Aristotle through Newton's abstract object and up to pseudo-vector in Einstein's relativity. Or, likewise, the evolution of the concept of "space" beginning from Aristotelean ether patch *via* Newton's infinite isotropic and homogeneous vacuum and Einstein's space-time up to Edward Witten's 10-dimensional mathematical object. It should be added that the ideas of Maxwell's 1877 paper were generalized and elaborated by a number of scholars (see, for instance [32 - 34]). As a result, the above mentioned epistemological "mature theory-change model" [29] was proposed (see also the last chapter). According to the epistemic model, the origins of scientific revolutions lie not in a clash of fundamental theories with facts, but of "old" fundamental theories with each other, leading to contradictions that can only be eliminated in a more general theory. The key role in theory change is played by the proponents of the old paradigms' ***dialogue*** that leads to mutual grinding and interpenetrations of the participants' views .

The very realization of reductionist and synthetic research programmes is brought about by the clash of mature theories which they are designed to eliminate. Having compared the heuristic potentials of the reductionist and the synthetic programmes, I favor the latter group since it has the following *objective* advantages. Firstly, synthetic programmes should provide a greater empirically-progressive shift of problems solved than the reductionist ones. Secondly, only these programmes can rationally explain the use of the so-called crossbred theoretical objects which spring from the coincident theories. For instance, if one considers the structure of two basic modern theories - quantum theory and general relativity - he finds that their global theoretical schemes arose from the unification of the crossbred theoretical schemes (see the third chapter for details).

Every case of different programmes' encounter leads to a situation when a multitude of *hybrid* models occurs formed by simple conjunctions from the models of different research programmes. However, the hybrid models appear to be self-contradictory; and when this is comprehended, the crossbreeds from the basic objects of all the cross-theories are constructed. A new mature theory is contrived due to crossbred domain growth.

This is not to derogate the role of experiments in science. On the contrary, the proposed epistemic model seems to elaborate further the point of view stated in the current philosophy of science literature that both theorists and experimentalists have breaks in their respective traditions, but they are not typically simultaneous [35, 36]. Theory development must have, to some extent, a life of its own. The development of two main cultures within science does not mean that the two do not speak to each other.

The epistemic model was illustrated with reference to physics in the early twentieth century, the three "old" theories in this case being Maxwellian electrodynamics, statistical mechanics and thermodynamics [37]. The world of "old", pre-Einsteinian physics was conceptually and socially patched. It was split on at least 3 research traditions belonging to maxwellian electrodynamics, classical thermodynamics and classical mechanics. Traditions organized around different isolated groups of phenomena generated little support for one another. The practitioners of each theoretical tradition acknowledged the existence of the other but went their own separate ways. With the advent of relativity and quantum theory, the conceptual unification of worldviews was accompanied by a social unification of practice.

In the process of interpenetration of the paradigms different theories begin to coincide with each other more and more. For instance, consider the blackbody radiation problem [37]. For the case of the radiation in a cavity the classical electrodynamics led to the so-called 'ultraviolet catastrophe', while classical thermodynamics (and statistical physics after it) insisted on the accomplishment of the first law of thermodynamics (energy conservation law).

The transformation of classical electrodynamics, when the electromagnetic field energy was taken in statistical mechanical way, *i.e.* when the energy of Planck's oscillators was quantized, had reconciled – in these local field – the consequences of electrodynamics with the consequences of thermodynamics and statistical physics.

Another startling example is reconciliation of General Relativity and Quantum Field Theory that led Stephen Hawking to his astonishing discovery of 'particle creation by the gravitation field of nonrotating black hole' [38].The introduction of the notions of gravitational field temperature and entropy was inescapable after the encounter of thermodynamics, statistical mechanics and Einstein's general relativity. Yet it led to the paradoxes and to blatant contradictions of one mature theory to another. And only when quantum field effects were taken into account and when it was exhibited that a black hole should create particles it became possible to reconcile the consequences of thermodynamics, statistical physics, general relativity, quantum field theory [39].

The next blatant example is the reconciliation by Newton of the Mundane physics (Huygens's pendulum experiments) with Divine mathematics (Kepler's laws) that successfully described the motion of the planets around the Sun [40]. In the long run, on revealing the cross-contradiction, a situation occurs when the consequences from different theories with different theoretical ontologies that describe the behavior of different crossbred objects begin to coincide. William Whewell, a polymath master of Trinity College, Cambridge, and Maxwell's teacher, had famously called it '*the coincidence of inductions*'.

> "That rules springing from remote and unconnected quarters should thus leap to the same point, can only arise from that being a point where truth resides" ([41], 65).

It is the growth (and even confluence) of the hybrid domains that constitutes the progress of scientific knowledge that shows itself up in grinding of the theories to each other, in reconciling and even unification of substantially different scientific theories. The crossbred objects and the relations between them form the condensations, the 'knots' of scientific knowledge that are conserved in theory change processes. The most apparent examples are the heliocentric system, the planetary model of atom, the vacuum model in quantum field theory in curved spacetimes. On the contrary, such bizarre theoretical objects as Ptolemy's concentric spheres around the Earth or luminiferous ether of the electromagnetic theory of light, absolute time or absolute space are lost in theory change since they ***hamper*** the unification process. They are too anthropocentric.

The typical example of "the daubing of untempered mortar elimination" principle for Maxwell was

> "The progress of science in Newton's time [which] consisted in getting rid of the celestial machinery with which generations of astronomers had encumbered the heavens, and thus 'sweeping cobwebs off the sky' "(Maxwell [42], 315).

Thus internal realism is subjective in a sense since it sanctions arbitrariness in choosing the ways of breaking the world up into different objects. Yet as a result of an encounter of two or more subjective ways of breaking up the world there appears such a new way that is more objective in respect to the old ones. The demarcation line between the theoretical ontologies becomes thinner. For instance, the Aristotelian physics strictly separated natural and violent motions, as well as the terrestrial (sublunary) and celestial (eternal) worlds. Newtonian theoretical ontology with its isotropic and homogeneous empty space was established only after the elimination of 55 spheres of Ptolemy cosmology conjured up to divide two Aristotelian worlds. Likewise, the superspace of supersymmetry theories was established only after eliminating the demarcation line between the bosons and the fermions.

In my judgement, one should discern internal and external facets (contexts) of the unification processes. For instance, one of the editors of the path-breaking volume "*The Disunity of Science*" – Peter Galison – admonishes that

> "modern talk of unification in the sciences originates in the German-speaking countries of the mid-19th century. For it was there, amidst the protracted political struggle for German unification, that scientific community was raised to a scientific-philosophical ideal" ([43], 3).

This deep intervening of political and scientific ideals is characteristic for Rudolf Virchow's "*Strivings Toward Unity*" (1847) and Hermann Helmholtz's pre-eminent "*The Aim and Progress of Physical Science*" (1869).

On the other hand, Rudolf Carnap's fascinating idea of creating of Unified Science sprung out of a cultural movement, and was not merely a specific addendum to the philosophy of language or philosophy of science. It reflected a post-war delusive hope that an international scientific worldview could deter the divisive racial and nationalistic worldviews.

Yet not only the political influence can be dominant. The unity of knowledge can be easily linked with monotheism. For instance, according to Ian Hacking ([44], 46), Michael Faraday's life (unification of electricity and magnetism) was located in the Sandamanian sect of upright protestants, as rigorous in their lives as they conceived God to be in arranging the Universe. Einstein's (unification of maxwellian electrodynamics and statistical thermodynamics, unification of special relativity and Newtonian gravity) cultural roots in Judaism were deep. Likewise, Abdus Salam's (unification of quantum electrodynamics and weak interactions theory) first Dirac lecture ends with an eloquent passage from the Quran.

In our times some new connotations have been affixed to unity and disunity. Instead of affiliation between unity and internationalism, liberal democracy, and a rational worldview, the axis unity/disunity has come to polarize around ideas of cultural autonomy in opposition to forces of homogenization, hierarchy and domination.

The incommensurability claim (that Newtonian mechanics and Einsteinian mechanics would not meet) found expression in the corresponding 'frameworks' of Rudolf Carnap, in the paradigms of Thomas S. Kuhn and in the conflicting interest-theoretical accounts of the Strong Programme of Sociology of Science (Barry Barnes and David Bloor). Against this tenet Peter Galison and Rinat Nugayev argue that the different paradigms, research programmes and research traditions work out ingenious "*local trading zones*" (Galison) or "*crossbred domains*" (Nugayev) in which they can successfully coordinate their practices.

Hence the stable totality of scientific knowledge, human culture and human Dasein is provided not by a situation when a "hundred flowers bloom" and a "hundred schools of thought contend" simply co-exist. And evidently not by a situation when one theoretical tradition suppresses the other ones (notorious cultural imperialism). The true stability is reached when the co-existing traditions *interact* and *interpenetrate* each other.

Just as Peter Galison [45] had convincingly demonstrated, the stable totality of scientific knowledge is provided not by theoretical traditions only, but by experimental traditions as well. He stressed that the content of experimentation has practically nothing to do with "observation" as a problem in the perception of instrument readings. In particular, apparently missing from the positivist account is the arena of argument situated between perception and the establishment of facts.

An experimentalist's main problem is to eliminate all the disturbing effects like gravitational field of the Earth fluctuations, *etc*. But in reality he can never assert that *all* the disturbing effects have been successfully eliminated. Hence between first suspicion and first argument there is a many-layered protracted process through which belief is progressively reinforced. It is clear that subtle interaction of theoretical and experimental traditions is far too complex to be parceled into a finite list of all possible epistemological backgrounds. However it is a humdrum that the "critical experiment" concept is an obvious over-simplification.

At first, as Pierre Duhem and Willard Quine have ascertained, the experiments refute not an isolated hypothesis but a *bundle* of "interconnected beliefs" (the so-called '*Duhem-Quine thesis*'). According to Quine,

> "My counter-suggestion, issuing essentially from Carnap's doctrine of the physical world in the *Aufbau*, is that our statements about the external world face the tribunal of sense experience not individually but only as a corporate body" ([46],p.344).

This is explained by the fact that

> "the totality of our so-called knowledge or beliefs, from the most casual matters of geography and history to the profoundest laws of atomic physics or even of pure mathematics and logic, is a man-made fabric which impinges on experience only along the edges. Or, to change the figure, total science is like a field of force whose

> boundary conditions are experiences. A conflict with experience at the periphery occasions readjustments in the interior of the field. Truth values have to be redistributed over some of our statements....But the total field is so undermined by its boundary conditions, experience, that there is much latitude of choice as to what statements to re-evaluate in the light of any contrary single experience" ([46], p.345).

At second, the reciprocal influence of theory on experiment turns out to be less subtle. It only commences with that the theoretical presuppositions carve out a piece of phenomenal world as interesting for a researcher. Peter Galison reveals the different levels of theory that exert their influence on the experiments. He turns to Fernan Braudel's meta-historical studies who, in his work on the Mediterranean civilization, found it useful to parse events into what he called a 'geographical time', a 'social time', and an 'individual time'. By comprehending the intersection of these geographical, social and individual constraints, Braudel provides a richer explanation of why certain decisions are made in history. Likewise Galison unfolds (1) long- term theoretical constraints (the conservation of energy), (2) the middle-level theoretical constraints (Holton's themes of unification, discreteness, continuity, *etc.*) and finally (3) particular theories and models.

The intersection of the above mentioned constraints (1) – (3) helps to provide a more subtle and complete comprehension of a history of science event.

Just as Renaissance society, for instance, embraced both high and low cultures, we can consider the discipline of physics as encompassing a culture of experimenters and one of theorists, each with its own standards of demonstration, commitments to methods, and programmatic goals.

Thus, Galison's final conclusion, which is of great significance for the construction of epistemological models and that I wholeheartedly accept, is the following one.

> "Both theorists and experimentalists have breaks in their respective traditions, but *they are not typically simultaneous*" ([45], p. 12; my italics).

5. A CASE STUDY: THE COPERNICAN REVOLUTION AND ELIMINATION OF GLARING CHASM BETWEEN MUNDANE PHYSICS AND DIVINE ASTRONOMY

As Imre Lakatos and Elie Zahar had vigorously maintained, both Ptolemy and Copernicus elaborated their genuine *research programmes* [47]. They did not merely test conjectures or try to put in order a vast multitude of observational data. Withal, both programmes branched off from the Pythagorean – Platonic one. The latter's pivotal tenet consisted in that, since heavenly bodies are perfect, *all astronomical appearances should be 'saved' by a combination of as few uniform circular motions as possible.* This basic principle remained the cornerstone of the *heuristic* of both programmes. Pythagorean-Platonic proto-programme contained no directives as to where the center of the universe lies, though.

In this case the '*heuristic*' turned out to be 'primary', while the '*hard core*' only 'secondary'. The geocentric hypothesis 'hardened' into a hard core assumption due to the impetuous advancement of Aristotelian physics, with its natural and violent motions and its strict separation of the terrestrial (sublunary) and celestial (eternal) worlds. Aristotelian 'physics' constituted an empirical discipline *par excellence.* It accumulated common experience much better than science of Galileo, Kepler and Newton. Indeed, hard bodies quite naturally fall down and fire soars up.

It is a platitude that inertial motion is not an experimental fact; and common experience evidently contradicts it. Moreover, everyone knows that "there is no void existing separately" (Aristotle [48]). According to our common experience, lucidly moulded on the categories of Aristotelian metaphysics, common reality in which one lives and acts (the '*lebenswelt*') is not a mathematical one. This kind of reality is the domain of uncertainty, unsteadiness, inaccuracy, comprehended by the words "almost", "a kind of", "more or less", "approximately", *etc.*

Nature is free of circles, ellipses, straight lines, plane surfaces and well swept streets.

Hence the Greek thought could not stand the tenet that exactness can exist in sublunary world and that "the matter of our sublunary world can represent mathematical entities" [49]. Rather, Aristotle asserted that "mathematical exactness should be demanded only for the objects lacking matter. Thus the mathematical way is not appropriate for natural science since Nature in all the cases is connected with Matter" [50].

Nevertheless, the divine world is quite the opposite. The heavens are made up of an entirely different immutable substance, the notorious ether (or 'the fifth element'). Heavenly bodies constitute spherical shells of ether. These shells fit tightly around each other in the following order: the shell of the Moon, Mercury, Venus, Sun, Mars, Jupiter, Saturn, and finally the shell of fixed stars. Each shell uniformly rotates, that accounts for the motion of the heavenly body contained in it. Outside the sphere of the fixed stars, there exists the prime mover that imports motion from the outside inward. The natural motions of heavenly bodies and their shells are perfectly circular and neither speeding nor slowing down, as it is required by their nature.

The perfect and ordered motions of the stars take place in full harmony with strict geometrical laws. Thus, due to Aristotle, *mathematical astronomy is blatantly possible and mathematical physics is certainly not.* So, the Greek astronomers not only effectively applied mathematics but diligently observed the sky with a help of subtle measuring devices. But they did not even attempt to mathematize terrestrial motions and to apply their devices to the phenomena on the surface of the Earth.

The summit of ancient science was reached by Claudius Ptolemy (87-150) – Hellenist astronomer, astrologer, mathematician, geographer (and even poet). His opus magnus – the superlative "*Almagest*" - dominated European thought for more than 14 centuries. (The book's title is derived from an Arabic "al-majasti" – the greatest; it's real title is "*Syntaxis Mathematica*"). Note that Ptolemy was a resident of Alexandria, the capital of Hellenized Egypt. Hellenistic civilization was a queer fusion of the Ancient Greek culture with that of Egypt and Babylon and an apparent departure from earlier Greek attitudes towards "barbarian" cultures. The extent to which genuinely hybrid Greco-Asian cultures emerged is contentious, but it is uncontroversial that the encounter and interaction of the three cultures captured the domain of mathematical astronomy [51].

Contrary to the Babylonians and Egyptians, who peered at the heavens to keep track of the seasons for practical reasons, the Greeks approached astronomy from a purely *theoretical* viewpoint: they wanted to *comprehend* the basic nature and makeup of the Universe. The Babylonian and Egyptian astronomy was contrasted with the Greek: the former were merely arithmetical and geometrical, but incomplete because lacking φυσϊόλόγια. *Ptolemy, as a true Hellenist, tried to skillfully balance himself between the three cultures.*

On the one hand, he was skeptical to heliocentric hypothesis of Aristarchus due to rational considerations grounded on the principles of Aristotelian physics. He easily admitted that it may perhaps be simpler, from a celestial viewpoint, to have the Earth spinning daily on its axis rather than the entire heavens rotating about the Earth. But this, he contended, fails to take into account the terrestrial physics.

For if the Earth moved,

> "Animals and other weights would be left hanging in the air , and the Earth would very quickly fall out of the heavens. Merely to conceive such things makes them appear ridiculous" (quoted from [52]).

On the other hand, since Aristotle was the only philosopher to whom Ptolemy referred explicitly, the author of the "*Almagest*" is often accused of slavish adherence to the tenets of Aristotelian philosophy. However, the things are not so simple both in physical and epistemological aspects. First of all, Ptolemaic ideal model of the solar system did considerably deviate from the orthodox Aristotelianism in a number of important facets.

a. Aristotle asserted that heavenly bodies should move in *single* uniform circles. Yet, in the Ptolemaic system, the motion of the planets is an artful *combination* of two circular motions. Moreover, at least one of them is *non-uniform*.
b. Aristotle also maintained that the Earth is located at the *exact* center of the Universe. But one should take into account that in the Ptolemaic system the Earth is slightly *displaced* from the exact center of the Universe. In Ptolemy's subtle cosmology there is no unique center of the Universe, since the orbit of the Sun and the planetary deferents all have slightly different geometric centers, none of which coincide with the Earth. And in the "*Almagest*" Ptolemy had frequently punctuated that the non-orthodox (from the Aristotelian stance) facets of his refined models all were *directly* dictated by observations.

Modern astronomy states, due to Kepler, that planetary orbits are actually ellipses. They possess two main hallmarks: (1) they are *eccentric*, *i.e.* the Sun is displaced from the geometric center of the orbit; (2) they are *elliptical*, *i.e.* the orbit is "elongated" along a certain axis. Nowadays Keplerian orbits are described by a quantity **e**, called *the eccentricity,* that measures their deviation from circularity. It can be demonstrated that the eccentricity of a Keplerian orbit scales as **e**, while the corresponding degree of elongation scales as $\mathbf{e}^2$. Since the orbits of the planets of solar system possess relatively small value of **e** ($\mathbf{e} < 0, 21$), it follows that, to a good approximation, these orbits can be taken as eccentric circles. One can evidently neglect the ellipticities of planetary orbits compared to their eccentricities. Namely that was done by Ptolemy in the "*Almagest*". "It follows that *Ptolemy's assumption that heavenly bodies move in circles is actually one of the main strengths of his model*" [53].

Ptolemy's second eminent hallmark was the notorious "*equant*". The latter constitutes a point near the center of the planet's orbit which, if you were to stand there and contemplate, the center of the planet's epicycle would always appear to move at the same speed. In accordance with Kepler's second law, the radius vector connecting a planet to the Sun famously sweeps out equal areas in equal times. In the approximation in which the planetary orbits are represented as eccentric circles, this law implies that a planet revolves around the Sun at a non-uniform rate. However, it can be exhibited that the non-uniform rotation of the radius-vector connecting the planet to the Sun implies a uniform rotation of the radius-vector connecting the planet to the so-called equant: *i.e.* the point directly opposite the Sun relative to the geometric center of the orbit. Ptolemy had discovered the equant quite empirically, and skillfully applied it in his ideal models to control the non-uniform rotation of the planets. As a matter of fact, "*this discovery is one of Ptolemy's claims to fame*" ([53], 6).

According to Imre Lakatos & Elie Zahar ([47], 371), after Eudoxus model (the system of rotating spheres) was abandoned, every move in the geostatic programme actually ran counter to the Platonic heuristic. The eccentric displaced the earth from the center of the circle. The Apollonian and Hipparchan epicycles meant that the real path of the planets about the earth was not circular; and finally the Ptolemaic equants entailed that even the motion of the epicycle's empty center was not simultaneously uniform and circular. The introduction of the equant was one of the heaviest blows upon the Platonic heuristic: it constituted its full elimination.

Thus, in the Ptolemaic research programme *mathematical exactness that demanded the introduction of noncircular orbits and the centers of rotation not coinciding with the earth center began to deviate more and more from the eminent principles of Aristotelian physics well-grounded empirically.* Hence one can evaluate *Ptolemaic cosmology as a* **dual** theoretical scheme *amalgamating the tenets of "Platonic mathematics" with those of "Aristotelian physics"*. (Just as, according to A. Einstein, Lorentz's eminent "*Theory of Electrons*" of the second half of the XIX century can be taken as a dual theoretical scheme amalgamating Newtonian mechanics with Maxwellian electrodynamics).

a. The realm of the blatant contradictions of Platonic mathematics and Aristotelian physics was the "*planet theory*" since it was namely the planet motions that needed for their description especially significant violations of the principles of Aristotelian physics. For instance, Proclus was interpreted by Plato in the "*Republic*" to be uttering that while the fixed stars are both regular and orderly, and sublunary things are both irregular and disorderly, the planets are intermediate between them [54].

b. The second domain of cross-contradictions turned out to be the *astrology* espoused in Ptolemy's "*Tetrabiblos*" (a work written after the "*Almagest*" but probably before the "*Planetary Hypotheses*"). In this book Ptolemy contemplates the interaction of divine and mundane worlds. He turns from pure contemplation of the celestial realm of ether to a meticulous investigation of the action of the heavens on the sub-lunar world famously engendered by the four elements.

> "A very few considerations would make it apparent to all that certain power emanating from the eternal ethereal substance is dispersed through and permeates the whole region about the earth, which throughout is subject to change, since, of the primary sub-lunar elements, fire and air are encompassed and changed by the motions in the ether, and in turn encompass and change all else, earth and water and the plants and animals therein" (Ptolemy [55], 1.2).

Yet there were substantial *frictions* between the basic hypotheses of the "*Almagest*" and the "*Tetrabiblos*". In both writings, the ethereal matter is simply a given. But the property of uniform circular motion in the "*Almagest*" is justified on purely *a priori* grounds (circular rotation being the only kind of eternally unchanging motion that can be comprehended), whereas the property of exerting a power of change on their elements is argued directly from empirical facts. Likewise, uniform circular motion is strictly mathematical behavior, which leads to the modeling of the "*Almagest*", while power to change the elements is by nature not mathematical. It operates with such qualities as "hot" and "cold", "wet" and "dry", *etc.*

> "And this posed a problem: how can cause-and-effect relations operating at the qualitative level, and largely within the 'irregular' sublunary part of the cosmos, be well described by the predictive mathematical models of astrology?" ([56], 35).

Yet Ptolemy neatly sweeps the problem away. Substantially he relies on the orderliness of the heavens to justify the mathematical structure of the predictive schemes of his astrology. However, he appeals to the disarray and complexity of the mundane environment to explain why astrological predictions, even concocted according to the most correct principles, are not certain to be borne out.

Hence after the translation of the "*Tetrabiblos*" into Latin (in the 12th century) Ptolemy's ingenious astrological scheme met severe criticisms that arose when the practice of predictions was comprehended to suggest "fatal necessity". The point turned out to be crucial to later integration into medieval Christian doctrine (Albertus Magnus and Thomas Aquinas), since the doctrine states that the individual sole must possess free will, in order to be responsible for its own choices and the consequences flowing from them. In particular, Gerard of Feltre's 13th century text "*Summa on the Stars*" considers the problem that astrological determinism created for the theological considerations:

> "If the stars make a man a murderer or a thief, then all the more it is the first cause, God, who does this, which it is shameful to suggest"(quoted from [57], 138).

From the vantage point of modern astronomy, Ptolemy's equant is an apparent precursor of Kepler ellipses. But for such stern critics as the Islamic astronomers and Copernicus the introduction of the equant was a typical "ad hoc$_3$ hypothesis" that blatantly contradicted the true spirit of Aristotle - Ptolemy programme – the principle of uniformity of motion in respect to the center of the Universe.

Ptolemy opted for the rational way of constructing subtle ideal models to mimic the peculiarities of planet rotations. The introduction of the '*equant*' into the positive heuristic of the geocentric programme was typical for all his shrewd constructions. Neatly combining epicycles and deferents, the author of the "*Almagest*", in accord with Eastern instrumentalist tradition, was (partly) inspired by a kind of "economy of thought" principle and initially did not bother himself with reflections on the essence of things.

Sometimes, describing the planetary motions, the author of the "*Almagest*" had simultaneously contrived several *alternative empirically-equivalent* theoretical schemes. Afterwards, as a rule, he opted for the mathematically simplest one. Ptolemy was especially skeptical to philosophical discussions and claimed that in astronomy one should choose the simplest mathematical model among those that are at his disposal.

For instance, in one of the well-known "*Almagest*" passages (III, ch.4), Ptolemy contends, in connection with his theory of the Sun, that the appearances may be saved on either an epicycle or on eccentric hypothesis. What Ptolemy actually says, after noting that both models can be used to account for the appearances in relation to the Sun, is that the eccentric hypothesis is to be preferred because "*it is simpler and effected by one, not two movements*". This eloquent passage enabled

Pierre Duhem to place Ptolemy among the ancient instrumentalists (Geminus, Ptolemy, Proclus, Simplicius). As he put it,

> "On the one side there was astronomy- geometers like Eudoxes and Callippus – formed mathematical theories by means of which the celestial motions could be described and predicted, while observers estimated to what degree the predictions resulting from calculations conformed to the natural phenomena. On the other side there was physics proper, or to speak in modern terms, celestial cosmology thinkers like Plato and Aristotle meditated on the nature of stars and the cause of their movements" (quoted from [58], 7). Shortly, the method of astronomers was evidently "instrumentalist", and the method of the physicists – "realist".

It seems an obvious over-simplification, though. In the "*Almagest*" one can find realist pieces as well, and in the "*Planetary Hypotheses*" Ptolemy holds an obvious *naïve realist* position. For instance, even in XIII, ch.2 of the "*Almagest*", when he asks the reader not to be dismayed by the complexity of the hypotheses that he has to use, his standpoint is **not** one of indifferences to the question of whether his devices represent the true system.

> "It is not fitting to compare human things with divine ones, nor to form beliefs concerning such great things from examples that are so unlike them. For what could be more unlike than those things that are eternal and unchanging and those that are never unchanging, or those that can be hindered by anything and those that cannot be hindered even by themselves?" ("*Almagest*", XIII, ch. 2, quoted from [54], 216).

For Ptolemy the mathematical agreement of the ideal models with the observational data constituted a pivotal evidence for the *physical reality* of the hypotheses, even if they seemed unduly complicated.

According to the "*Almagest*" (XIII.2), "one should try, as far as possible, to fit the simpler hypotheses to the heavenly motions, but if this does not succeed, [one should apply such hypotheses] which do fit. For provided that each of the phenomena is duly saved by the hypotheses, why should anyone think it strange that such complications can characterize the motions of the heavens when their nature is such as to afford no hindrance, but of a kind to yield and give way to the natural motions of each part, even if [the motions] are opposed to one another" (quoted from [58], 8).

Thus, all in all, in the "*Almagest*" Ptolemy holds an ***eclectic*** epistemological position, skillfully combining instrumentalist and naïve realist stances in the true spirit of Hellenistic culture. However, to grasp the dominating epistemological inclinations one has to listen to the other outstanding eclectic, Albert Einstein, who stubbornly tried to amalgamate Mach's positivism with Planck's and Boltzmann's realism. At the very outset of his pre-eminent paper "*On the Method of Theoretical Physics*" he confessed:

> "If you want to find out anything from the theoretical physicists about the methods they use, I advise you to stick closely to one principle: don't listen to their words, fix your attention on their *deeds*" ([59], 30; my italics).

So, if one takes the "*Almagest*" from this standpoint, one finds out that the most important results were obtained there on the basis of the "principle of economy of thought". In his research practice, his actual modus operandi Ptolemy acted as a committed instrumentalist. Hence, one should distinguish Ptolemy$_1$ – an instrumentalist and the author of "*Almagest*", and Ptolemy$_2$ – a realist and the author of "*Planetary Hypotheses*", "*Tetrabiblos*" and "*Optics*". It is only Ptolemy$_1$ who is usually associated with the slogan "saving the phenomena".

This is not to ignore Ptolemy's realist standpoint, since, after the "*Almagest*" he had to make the ends meet with the Zeitgeist, with the realist paradigm. To retain the achievements of the Aristotelian natural philosophy and to retain the achievements of his own theoretical astronomy, Ptolemy had to take the next step in elaborating the Aristotelian dualism of mundane and celestial phenomena in his "*Planetary Hypotheses*". Since the "*Almagest*" confines itself to mathematical models, Ptolemy had to try to provide preliminary physical models for the same constructions.

The "*Planetary Hypotheses*" book, only parts of which have become well known in the West recently, consists of the two volumes. In the first one Ptolemy corrects some of the parameters of the "*Almagest*" and suggests an improved model to explain planetary latitude. However, the second volume, dealing with ether hypotheses, added nothing new to the empirical success of the Ptolemaic programme, and can be taken as a multitude of ad hoc hypothesis.

Ptolemy intended to advance a celestial mechanism grounded in the mathematical theories set up in the "*Almagest*". His ideal models should "adhere to the principles that correspond to the continuous substance", that is, the ether. So, he orderly lists the characteristics of ethereal bodies: each is invariable and finite, receives no agitation, moves uniformly and circularly, and performs only a single, 'voluntary' motion. It is crucial that Ptolemy also claims that celestial bodies maintain a faculty which may be compared to the human faculties of vision and intelligence.

In his lucid account of the motion of celestial bodies, Ptolemy draws a sharp distinction between the Earth and the Heavens. His cosmology requires all the bodies, both celestial and terrestrial, to participate in the 'natural' motions of their own. Whereas an earthly body can be propelled violently with a motion not proper to it, a celestial one enjoys only uniform and circular motion. And, unlike sub-lunar motion, the motion of the heavens cannot be determined by outside forces. Ptolemy sharply criticizes Aristotle for using terrestrial examples to generate an image of the nature of the celestial spheres. The philosophers fail to understand planetary motion, he contends, since they "begin their analogy from the spherical motions that are near us", by using the motion of mundane bodies to comprehend celestial rotation. But Ptolemy's celestial spheres follow the laws that differ from those that govern terrestrial motion.

He also applies the terrestrial examples to explain the structure of the heavens. He takes the celestial spheres as parts of "universal animal". Birds and clouds may be compared to the planets since they are composed of the same elements that surround them, although their colors and densities are different from those of the air.

Ptolemy inveighs against the Aristotelian idea of the 'prime mover'. Like Aristotle, he admits that motion is transmitted by contact, yet the prime mover touches only the outermost of all the celestial bodies. Only the outermost sphere would rotate with its proper motion, since it could hardly transmit its own particular motion to the other celestial bodies.

To comprehend the order of the Universe, Ptolemy turns to another genuinely metaphysical concept, *the soul*. He readily admits that "the planets are endowed with a soul and are moved with a voluntary motion". It is the soul that engenders the faculty for producing brightness and motion. Such a faculty is natural to ether. The wandering stars command a substantially larger amount of the faculty than the other ethereal bodies due to the following reason.

Each planet has the power to direct not only its own motion, but the motions of the neighboring celestial bodies as well. To illustrate this faculty, Ptolemy applies for the terrestrial example of a very subtle kind: the motion of birds in the layer of air above the Earth. In sub-lunar world the psychic faculty of the bird sends an emission to its nerves and then its wings; it allows a bird to perform its movement. Similarly, the planet's psychic faculty sends emissions to the epicycle, then to the deferent, then to the outermost of that planet's celestial bodies. The emission functions as a "green light" (Murschel) to which each body responds with its own, 'voluntary' motion, just as the motion of the nerves differs from the motion of the wings.

The ideal model recapitulated should offer an answer to another basic question: how celestial bodies that are spherical by necessity connect to one another?; By what means do they transmit motion to one another? Citing the poles that Aristotle retained in his system, Ptolemy wants to know "Why isn't anything destroyed at the contact between the inner sphere and the first outer sphere? And why aren't all the motion[s] of the spheres equal in speed?"(quoted from [60], 39).

Since the ether is frictionless and nothing can slow down the uniformity of its movement, any supra-lunar, physical, fixed poles must be unable to impose motion on the surrounding ether. Furthermore, no sphere would need a mechanism to cause its motion since ether by nature will rotate uniformly.

Yet the poles need not be fixed, physical bodies. Hence in the true spirit of the "*Almagest*", Ptolemy takes this construct *mathematically*, as the points on a rotating sphere that do not trace out a circular path. However, this definition fails to satisfy him. Ptolemy finds it unrealistic that a mathematical object such as a point would be able to move a physical object like an ethereal sphere. Thus he contends that the now-physical points, also made of ether like the spheres they connect, could now function as movers for the spheres, since they would be nothing more than "knots" or "warts". And if they differ from the surrounding ethereal surfaces, they would "sink to the center of the world".

In constructing the abstract theoretical objects of his cosmology Ptolemy turns now to the quasi-Platonic "principle of economy of nature" instead of the instrumentalist "principle of economy of thought": "it is appropriate that we do [not] assume that anything be made in nature that has no meaning and that is of no use". In that respect Ptolemy's philosophical evolution strongly resembles the biography of Albert Einstein.

The question still remained as to whether Ptolemy's models represented *true* physical hypotheses for the cosmos. But almost all the "*Almagest*" geometrical models were reproduced in a comparable physical form. Moreover, the principles governing the heavens, that Ptolemy had accepted as necessary and true, were not compromised by his physical models. For instance, sawn-off pieces seemed no more unlikely than the ethereal shells first assumed to fill the heavens. These structures may seem mathematically trivial, but Ptolemy's object was to provide a suitable and consistent physical representation of the heavens, and not to invent new planetary theories.

Nevertheless,

> "the pronouncements in the "*Planetary Hypotheses*" are more equivocal than those of the "*Almagest*". Ptolemy is quite sure of the ethereal composition of the heavens, and also quite sure of the fundamental geometrical structures of the celestial motions; but the specific way that these geometrical structures are embodied in the ethereal matter is *open to alternatives* (complete spheres or equatorial slices, spinning driven by cosmic souls or by planetary rays or by the axes)" ([56], 35).

Likewise, no rigid links connecting the mathematical and the physical models were provided by Ptolemy, so that each new mathematical result would buttress the physical hypothesis and *vice versa.*

Thus, *after the "Planetary Hypotheses"the gap between the physical and mathematical principles became even deeper*, so Ptolemy's cosmology came under repeated attack during the Middle Ages [61]. ***It obviously contradicted the principles of monotheism not admitting the profound gap between the celestial and mundane worlds.*** In the second book of "*Planetary Hypotheses*", describing a physical actualization of the mathematical models of the planets in the "*Almagest*", the conflict with Aristotelian physics became especially sharp. It was in attempting to remove the discrepancies that the "School of Maragha" and also Ibn-Shatir in the 13th and 14th centuries invented new planetary models that anticipated Copernicus [62]. The Middle Age Arabic astronomers aimed at readjusting their science to become more scientifically coherent.

The new discipline of Nay'a (theoretical astronomy) was created due to the flourishing activities at the time of the early Abbasides that triggered a recovery of the sciences of antiquity with a deep desire to deploy them for the purposes of the time, a phenomenon that was not to be repeated until the time of the late European Renaissance. Note that once they could shun the discipline of astrology, then the importers of Greek astronomy, as well as the composers of Nay'a texts could pose as allies of the religious establishment. Most of the Arabic scientists were *religious functionaries* as well. Religion's main impact on astronomy was to force its separation from astrology and to compel attempts to reconcile Ptolemaic cosmology and Aristotelean physics [63]. Theology compelled astronomers to deny the divine attributes of the objects of the Universe, and so their motions had to be comprehended according to the mathematical laws of their models.

The re-consideration of Ptolemaic astronomy reached its full maturity during the 11th century. The most sophisticated attack on it was leveled by Ibn al-Haytham (965-1040), also known as Alhazen. He aimed at "*reconciliation of the mathematical and physical facets of astronomy*".

In "*Doubts Concerning Ptolemy*", published between 1025 and 1028, Alhazen deprecated the "*Almagest*", "*Planetary Hypotheses*" and "*Optics*", punctuating "innumerous contradictions" he uncovered there. For instance, Ibn-al-Haitham stressed that Ptolemy had set himself the task of accounting for the phenomena using uniform circular motions and that, since he had introduced the equant mechanism, he could not be considered to have succeeded. Ibn-al-Haitham objected to Ptolemy's lunar theory because it involved an imaginary point opposite the center of the deferent controlling the motion of the lunar apogee. This turned out to be physically impossible. Above all, he contended, astronomy should hinge upon real bodies, and not imaginary ones. As he put it,

> "Thus the duty of the man who investigates the writings of scientists, if learning the truth is his goal, is to make himself an enemy of all that he reads, and, applying his mind to the core and margins of its content, attack it from every side... I constantly sought knowledge and truth, and it became my belief that for gaining access to the effulgence and closeness to God, there is no better way that of searching for truth and knowledge" (quoted from [63], 72).

The Persian Abu Ubayad al-Juzjani contrived his own set of non-Ptolemaic models. Al-Juzjani's attempts were ultimately unsuccessful, but the search for alternate models continued during the twelfth century in the Islamic west. Astronomers of Andalusian Spain renounced geometric models that failed to satisfy Aristotle's axiom of uniform circular motion about a fixed point.

Eventually astronomers from Damascus and Samarkand had initiated the process later called the "Maragha Revolution" (the Maragha school's revolution against Ptolemaic astronomy). The astronomers of Maragha school were more successful than their Andalusian predecessors in contriving non-Ptolemaic configurations which eliminated the equant and eccentrics.

The pivotal aspect of the Maragha revolution included the realization that astronomy should aim to describe the behavior of physical bodies in mathematical language. But it should not remain a mathematical hypothesis only, which would merely "save the phenomena". Unlike the ancient Greek and Hellenistic astronomers (Eudoxes and Ptolemy, for instance) who were not concerned with the coherence between the mathematical and physical principles of a planetary theory, Islamic astronomers insisted on the need to match the mathematics with the real world surrounding them. The Maragha School gradually evolved from Aristotelian physics to a specific blend of empirical and mathematical methods after the work of Ibn al-Shatir. The hallmark of the Maragha Revolution was thus the shift away both from the philosophical foundations of Aristotelian physics and Ptolemaic astronomy and towards a greater emphasis on the empirical observation and mathematization of nature, as exemplified in the works of Ibn al-Shatir, Qushji, al-Birjandi and al-Khafri.

To recapitulate, the main purpose of all the Arabic theorists – whose work had a direct bearing on the theories of Copernicus – was to try to square the requirements of Aristotelian natural philosophy with the mathematical models of Ptolemaic cosmology. Only in very peculiar cases were objections to Greek astronomy made on the basis of its failure to account for the observed facts.

The Golden Age of Islamic astronomy had extended from the mid-thirteenth to mid-fourteenth century. During this time, strenuous attempts were made to reform Ptolemaic astronomy by Nasir al-Din al-Tusi (1201-1274), Mu'ajjad al-Din al-Urdi (died in 1266), Qutb al-Din al-Shirazi (1236-1311), Ibn al-Shatir (1305-1375), and others.

In his path-breaking "*Memoir on Astronomy*" Nasir al-Din al-Tusi (1201-1274) considered the celestial entities as moving physical objects rather than abstract geometric 'kinetic' constructs. To object the violation of the principle of uniform circularity in the "*Almagest*" he had famously invented a device sometimes known as the "Tusi couple".

To recapitulate, the works of Islamic astronomers are necessary for rational reconstruction of Ptolemy-Copernicus transition since they constituted important momenta of general trend that reached its climax in Copernicus's "*De Revolutionibus*" – to "*reconcilethe mathematical and physical aspects of astronomy*".

The pivotal contradiction awareness and elimination of which constitutes an intertheoretic context of the Copernican revolution was excavated in the middle of the XX-th century by French philosopher and historian of science Alexander Koyré. Digging still deeper into the background, he unfolded the "*tremendous gap*" between mathematical astronomy and Aristotelian qualitative physics. Hence the heliocentric programme did not spring out of the common discrepancies between the Ptolemaic cosmology and experience. Rather, Copernicus was inspired by aesthetic and metaphysical – *and strictly theological in essence* - considerations evidently connected with elimination of the abovementioned dualism.

In the pre-eminent Introduction to his "*De revolutionibus orbium coelestium*", published in 1543 in Nuremberg and ostensibly dedicated to "his holiness Pope Paul III", Nicolaus Copernicus, a canon at Frombork Cathedral who's maternal uncle was Lucas Watzenrode, the bishop of Warnia, directly emphasizes that

> "I was impelled to consider a different system of deducing the motions of the universe's spheres for no other reason than the realization that *astronomers do not agree among themselves in their investigations of this subject.*"

For, in the first place, they are so uncertain about the motion of the sun and moon that they cannot establish and observe a constant length even for the tropical year. Secondly, in determining the motions not only of these bodies but also of the other five planets, they do not use the same principles, assumptions, and explanations of the apparent revolutions and motions. For while some employ only homocentrics, others utilize eccentrics and epicycles, and yet they do not quite reach their goal. For although those who put their faith in homocentrics showed that some nonuniform motions could be compounded in this way, nevertheless by this means they were unable to obtain any incontrovertible result in absolute agreement with the phenomena. On the other hand, those who devised the eccentrics seem thereby in large measure to have solved the problem of the apparent motions with appropriate calculations. But meanwhile they introduced a good many ideas which apparently *contradict the first principles of uniform motion.* Nor could they elicit or deduce from the eccentrics the principal consideration, that is, the structure of the universe and the true symmetry of its parts" (Copernicus [64]; my italics).

As a consequence,

> " I began to be annoyed that the movements of the world machine, *created for our sake by the best and most systematic Artisan of all,* were not understood with greater certainty by the philosophers, who otherwise examined so precisely the most insignificant trifles of this

world".

The crux of the paradox, as seen by Copernicus, consists in non-ideal movement of the planets. Nevertheless, they necessarily belong to ideal spheres and *should* be engaged in uniform motions along the circles or along their combinations. Like Aristotle, Copernicus was convinced that the supposed perfection of heavens requires celestial bodies to execute uniform circular motion only. Hence he was spurred to reject Ptolemy's pagan notorious equant model.

Inspired by the "*best intentions to demonstrate the Divine Order of the Heavens*", Copernicus proposes to place the center of the Universe on Sun. However namely this causes the deep paradoxes within the Aristotelian physics caused with the notions of 'natural' and 'violent' movements.

As a matter of fact Copernicus had contrived a genuine *crossbred* theory (analogous to first Planck's 1900 semiclassical theory) that paved the way to divine mathematics and mundane physics *interaction* and *interpenetration*. As modern French historian put it

> "Copernicus in an insinuating manner and maybe unconsciously had introduced into Aristotelian fortress two small assumptions with the help of which Kepler, Galileo and Descartes blowed it up" ([65], 430).

It's no surprise that Copernicus appeals to a favorable clergy consisting of Pope Paul III (to whom his "*De Revolutionibus*" was dedicated), Pope Climentus VII (who adopted his opus magnus and even insisted on its publication), Nicholas Schonberg, cardinal of Capua, his maternal uncle, bishop of Warmia, Tiedemann Giese, bishop of Chelmno, and so on, and deprecates Ptolemy for *paganism*. He batters Ptolemy for that his sophisticated system containing tens and hundreds of epicycles and equants is nevertheless lacking single, monotheistic God, 'the best and most systematic Artisan'; so the various parts of Ptolemaic system reflect the various plans of different pagan "artisans". Thus,

> "on the contrary, their [Ptolemy's proponents] experience was just like someone taking from various places hands, feet, a head, and other pieces, very well depicted, it may be, but not for the representation of a single person; since these fragments would not belong to one another at all, a monster rather than a man would be put

together from them" (introduction to [64]).

Namely that way Copernicus (maybe even unconsciously) paved the way to Galileo. If there is no strict demarcation line between Divine and Mundane worlds and the Earth is just an ordinary planet, the mathematical principles and methods should be applicable to its rotations around its own axis and around the Sun, as well as and to the bodies moving along its surface. In future in Galileo's writings Aristotelian 'natural' movements turned out to be Newton's 'inertial' ones.

For more sweeping and systematic rational reconstruction of Copernican programme 'hard core', 'heuristic' and 'protecting belt' one has to turn to the works of Nicolas of Cusa (1401-1464) who was a cardinal of the Catholic church from Germany. He was a philosopher, theologian, jurist, mathematician and an astronomer, one of the greatest polymaths of the XVth century. Nicolaus was widely read. In particular, Nicolaus Copernicus, Johannes Kepler and Galileo Galilei were all aware of his writings. In "*De Concordantia Catholica*", "*De Docta Ignorantia*" and "*De Visione Dei*" the metaphysical intuitions constituting the "Zeitgeist" were systematically ordered by Cusanus. Nicolas's 'monotheistic creationism' was directed against the cosmos of Aristotle and Ptolemy: as a result of divine creation skies should not differ significantly from the Earth. On the other hand, Man was treated as a "second God" and hence he was an architect of ideal (mathematical) theoretical objects. This paved the way to mathematical physics.

Galileo descends mathematics from the Skies to the Earth being inspired by Copernicus and Plato (especially his dialogue "*Timaeus*") and by his own astronomical observations as well. In the especially pithy programmatic work "*The Assayer*" (1623) Galileo maintains that

> "philosophy is written in that great book which ever lies before our eyes — I mean the universe — but we cannot understand it if we do not first learn the language and grasp the symbols, in which it is written. This book is written in the mathematical language, and the symbols are triangles, circles and other geometrical figures, without whose help it is impossible to comprehend a single word of it; without which one wanders in vain through a dark labyrinth" (Galilei [1623],1957, quoted from [66],75).

And to assure the reader by whom the Book was written, in his path-breaking introduction to the "*Dialogue Concerning the Two Chief Systems – Ptolemaic and*

Copernican" Galileo stressed that

> "he who looks the higher is the more highly distinguished, and turning over the great book of nature (which is the proper object of philosophy) is the way to elevate one's gaze. And though whatever we read in that book is the creation of the omnipotent Craftsman, and is accordingly excellently proportioned, nevertheless that part is most suitable and most worthy which makes His work and His craftsmanship most evident to our view" [67]. For Galileo "*mathematics is the language with which God has written the Universe*" (quoted from [68], 2).

Galileo's interpretation of Christian theology is famously inspired by Plato and his "*Timaeus*" in particular; especially influential was a myth of creation of the Universe. The central figure of the dialogue – the Demiurge, a divine Craftsman – imposes mathematical order on a preexistent chaos to contrive the ordered Universe ('Cosmos'). For that purpose he cuts out small triangles to construct four regular solids; and then he applies these solids for architecting real bodies, plants and animals out of them (Plato [69]). Furthermore, it was namely in "*Timaeus*" where the notion of a divine Craftsman was enriched by the notion of pre-established itinerary devised by him.

Thus Nature is empty and ordered since in the process of its creation God put strict mathematical necessity in it. Hence the Mathematical Knowledge is not simply true but is sacred even more than the Bible. While there are plenty interpretations of the Bible, the Mathematical Truths are unique and should be necessarily kept out of discussions.

At the sake of mathematization Galileo radically transforms the methodology of natural science elevating idealization and thought experiments up to the ranks of leading methods of science [70]. All that made it possible for Galileo to enunciate "the principle of inertia" and to come close to Newton's second law of dynamics [71].

Likewise Platonic (and Neo-Platonic) motives and especially "*delightful accordance between the Cosmos and the Holy Trinity*" (Kepler [72]) inevitably brought Johannes Kepler to the search for mathematical laws governing the planet motions. Properly educated in the Lutheran faith, Kepler devoted himself to the "Book of Nature", taking his work in the realm of natural science as the service of a priest. "We astronomers are priests of the highest God in regard to the book of Nature" (quoted from [73]).

However, there existed an essential difference between the views of Kepler and Copernicus. In Copernicus's theory the planetary motions were circular. Hence the Sun was not the center of force and its place should coincide with the Earth orbit center. Only Kepler's reflections on the source of planetary motion revealed the role of the Sun and it helped him to contrive the mechanics of planetary motion (see, for instance [74]).

Johannes Kepler took the Sun as the geometric and dynamical center of the Cosmos on the grounds of Trinitarian theology. The coequality of Father, Son, and Spirit required the continuity of the center, periphery and space of the Cosmos. The Sun was God the Father, the Stars God the Son and the planets were Holy Ghost. Likewise, Kepler's stubborn attempts to reveal a universal law that would account for the motion of both the Earth and the planets were also inspired by the pivotal analogy between the created Cosmos and the uncreated Trinity.

Johannes Kepler took the second step towards the unification of mathematical astronomy and empirical physics and uncovered the laws that roughly broke the Aristotle-Ptolemy principle of uniform rotation of divine bodies. Kepler's three laws turned out to be the first scientific laws taking mathematical form. Skies started to destroy the qualitative physics. Unification of the divine and of sublunary worlds moved Aristotelian qualitative obsolete physics aside (Kepler [75]).

Sir Isaac Newton's aimed at discovering the laws that govern the motion of both terrestrial and divine motions. The first milestone consisted in demonstration, with the help of Copernicus and Galileo heuristic, that it was the same force that attracted all the bodies to earth that forced the moon to orbit the Earth. Namely that was done in "*Philosophiae Naturalis Principia Mathematica*" (1687). As is stressed in Mr. Roger Cotes's preface

> "Therefore it is plain that the centripetal force, by which the Moon is perpetually, either impelled or attracted out of the tangent and retained in its orbit, is the very force of terrestrial gravity reaching up to the Moon" (Newton [76], 31).

Newton generated the 'hard core' of his research programme *via* unifying the theoretical schemes of Copernicus, Kepler, Hook and Galileo and coming up to the conjunction of three laws of dynamics with the law of gravitation. All these enabled him to augment the empirical content of the Copernican research programme.

6. CONCLUDING REMARKS

The various arguments pro et contra convergent realism are scrutinized. It is argued that to overcome the most apparent drawbacks of convergent realism one has to turn from classical or 'metaphysical' realism to non-classical or 'internal' one, grounded on Kantian epistemology, and correspondingly to the so-called "coherent theory of truth". But then the question necessarily arises: where does the objectivity of science comes from?

One of the answers is provided by the author's lucid epistemological model of mature theory change. According to the model, the origins of scientific revolutions lie not in a clash of fundamental theories with facts, but of "old" mature research traditions with each other, leading to contradictions that can only be attenuated and even eliminated in a more general theoretical approach. The model is illustrated and elucidated with reference to the Copernican Revolution. It is exhibited that this revolution consisted in eliminating (in the writings of Copernicus, Kepler, Galileo and Newton) the profound gap between Mundane physics and Divine astronomy meticulously unfolded in the writings of the Middle Age Arab- Islamic theorists and astronomers.

REFERENCES

[1] D. Papineau, Introduction.*The Philosophy of Science.* Oxford University Press: Oxford, 1996, pp. 3-22.

[2] B.C. Van Fraassen, *The Scientific Image.* Clarendon Press: Oxford, 1980. [http://dx.doi.org/10.1093/0198244274.001.0001]

[3] L. Laudan, "A Confutation of Convergent Realism." *In: The Philosophy of Science.* Oxford University Press: Oxford, 1996, pp. 107-138.

[4] T.S. Kuhn, "Newton's Optical Papers." *In: Isaac Newton's Papers and Letters on Natural Philosophy.* Harvard University Press: Cambridge, Massachusetts, 1958, pp. 27-45.

[5] T.S. Kuhn, "The Function of Dogma in Scientific Research." *In: Scientific Change.* Harvard University Press: Cambridge, Massachusetts, 1963, pp. 347-359.

[6] T.S. Kuhn, "Objectivity, Value Judgement and Theory Choice." *In: The Essential Tension.* University of Chicago Press: Chicago, 1973, pp. 320-339.

[7] J. Worrall, "Fresnel, Poisson and the white spot: the role of successful predictions in the acceptance of scientific theories." *In: The Uses of Experiment.* Cambridge University Press: Cambridge, 1989, pp. 135-157.

[8] D. Richard, *String Theory and the Scientific Method.* Cambridge University Press: Cambridge, 2013.

[9] Ya.B. Zel'dovich, and I.D. Novikov, *Stars and Relativity.* Pergamon Press: London, 1973.

[10] S. Zizek, "Introduction." *In: Mapping Ideology.* Verso: London, 2001, pp. 3-20.

[11] N. Cartwright, *How the Laws of Physics Lie.* Cambridge University Press: Cambridge, 1985.

[12] I. Hacking, *Representing and Intervening.* Cambridge University Press: Cambridge, 1985.

[13] J. Van Dongen, *Einstein's Unification.* Cambridge University Press: Cambridge, 2010. [http://dx.doi.org/10.1017/CBO9780511781377]

[14] J. Worrall, "Structural Realism: The Best of Both Worlds?" *In: The Philosophy of Science.* Oxford University Press: Oxford, 1996, pp. 100-124.

[15] I. Kant, *The Critique of Pure Reason.,* P. Guyer, A.W. Wood, Eds., Cambridge University Press: Cambridge, 1998. [http://dx.doi.org/10.1017/CBO9780511804649]

[16] H. Poincaré, *Science and Hypothesis.* Flammarion: Paris, 1906.

[17] S. Saunders, "Decoherence, relative states and evolutionary adaptation", *Found. Phys.,* vol. 23, no. 12, pp. 1553-1585, 1993. [http://dx.doi.org/10.1007/BF00732365]

[18] C. Cutler, and R.M. Wald, "Existence of radiating Einstein-Maxwell solutions", *Class. Quantum Gravity,* vol. 6, no. 4, pp. 453-466, 1989. [http://dx.doi.org/10.1088/0264-9381/6/4/006]

[19] S. Psillos, "Is structural realism the best of both worlds?", *Dialectica,* vol. 49, pp. 15-46, 1995. [http://dx.doi.org/10.1111/j.1746-8361.1995.tb00113.x]

[20] R. Frigg, and I. Votsis, "Everything you always wanted to know about structural realism but were afraid to ask", *Eur. J. Philos. Sci.,* vol. 1, no. 2, pp. 227-276, 2011. [http://dx.doi.org/10.1007/s13194-011-0025-7]

[21] V.S. Stepin, *Theoretical Knowledge.* Springer: Dordrecht, 2005. [http://dx.doi.org/10.1007/1-4020-3046-0]

[22] H. Putnam, "Two Philosophical Perspectives." *In: The Nature of Truth. Classic and Contemporary Perspectives.* The MIT Press: Cambridge, Massachusetts, 2001, pp. 251-258.

[23] M. Lynch, Ed., *The Nature of Truth. Classic and Contemporary Perspectives.* The MIT Press: Cambridge, Massachusetts, 2001.

[24] R.C. Walker, "The Coherence Theory." *In: The Nature of Truth. Classic and Contemporary Perspectives.* The MIT Press: Cambridge, Massachusetts, 2001, pp. 123-158.

[25] C.S. Pierce, "How to Make Our Ideas Clear." *In: The Nature of Truth. Classic and Contemporary Perspectives.* The MIT Press: Cambridge, 2001, pp. 193-210.

[26] B. Blanshard, "Coherence as the Nature of Truth." *In: The Nature of Truth. Classic and Contemporary Perspectives.* The MIT Press: Cambridge, 2001, pp. 103-122.

[27] L.M. Alcoff, "The Case for Coherence." *In: The Nature of Truth. Classic and Contemporary Perspectives.* The MIT Press: Cambridge, 2001, pp. 159-182.

[28] M. Foucault, "Truth and Power." *In: The Nature of Truth. Classic and Contemporary Perspectives.* The MIT Press: Cambridge, 2001, pp. 317-320.

[29] R.M. Nugayev, *Reconstruction of Mature Theory Change: A Theory –Change Model.* Peter Lang: Frankfurt am Main, 1999.

[30] N. Maxwell, "Unification and revolution: A paradigm for paradigms", *J. Gen. Philos. Sci.,* vol. 45, pp. 133-149, 2014. [http://dx.doi.org/10.1007/s10838-013-9225-5]

[31] J.C. Maxwell, "Hermann Ludwig Ferdinand Helmholtz", *Nature,* vol. 15, pp. 389-391, 1952.

[32] L. Tisza, "The Conceptual Structure of Physics", *Rev. Mod. Phys.,* vol. 35, pp. 151-18, 1963. [http://dx.doi.org/10.1103/RevModPhys.35.151]

[33] M.I. Podgoretzky, and J.A. Smorodinsky, On the Axiomatic Structure of Physical Theories.*The Physical Theory.* Nauka: Moscow, 1980, pp. 53-61. (in Russian)

[34] C. Rovelli, *Quantum Gravity.* Cambridge University Press: Cambridge, 2008.

[35] A. Pickering, *Constructing Quarks. A Sociological History of Particle Physics.* The University of Chicago Press: Chicago, 1985.

[36] P. Galison, *How Experiments End.* The University of Chicago Press: Chicago, 1987.

[37] R.M. Nugayev, "Why did new physics force out the old?", *Int. Stud. Philos. Sci.,* vol. 10, no. 2, pp. 127-140, 1996.
[http://dx.doi.org/10.1080/02698599608573534]

[38] S.W. Hawking, "Particle Creation by Black Holes", *Commun. Math. Phys.,* vol. 43, no. 3, pp. 199-220, 1975.
[http://dx.doi.org/10.1007/BF02345020]

[39] B.S. De Witt, "Quantum-Field Theory in Curved Spacetimes", *Phys. Rep.,* vol. C19, pp. 297-357, 1975.

[40] R.M. Nugayev, "The Ptolemy-Copernicus Transition: Intertheoretic Context", *Almagest,* vol. IV, no. 1, pp. 96-119, 2013.
[http://dx.doi.org/10.1484/J.ALMA.5.100985]

[41] W. Whewell, *The Philosophy of the Inductive Sciences, founded upon their history.* vol. 2. John L. Parker: London, 1847.

[42] J.C. Maxwell, *A Treatise on Electricity and Magnetism.* 3rd ed. Dover: New York, 1954.

[43] P. Galison, and D. Stump, *The Disunity of Science.* Stanford University Press: Stanford, 1996.

[44] I. Hacking, "The Disunities of Science." *In: The Disunity of Science.* Stanford University Press: Stanford, 1996, pp. 37-74.

[45] P. Galison, *How Experiments End.* The University of Chicago Press: Chicago, 1987.

[46] W.V. Quine, "Two Dogmas of Empiricism." *In: Contemporary Analytic and Linguistic Philosophies.* Prometheus Books: N.Y., 2000, pp. 331-349.

[47] I. Lakatos, and E. Zahar, "Why did Copernicus's Research Program Supersede Ptolemy's?" *In: The Copernican Achievement.* University of California Press: Los Angeles, 1974, pp. 168-192.

[48] Aristotle, *Physics* Cambridge,MA: Harvard University Press, 2007.

[49] A. Koyre, *From the Closed World to the Infinite Universe.* John Hopkins Press: Baltimore, 1957.

[50] Aristotle, *Metaphysics* Prometheus Books: NY, 1991.

[51] O. Neugebauer, *A History of Ancient Mathematical Astronomy.* Springer: Berlin, 1975.
[http://dx.doi.org/10.1007/978-3-642-61910-6]

[52] M. Kline, *Mathematics and the Search for Knowledge.* Oxford University Press: Oxford, 1986.

[53] R. Fitzpatrick, *A Modern Almagest. An Updated Version of Ptolemy's Model of the Solar System.* Lulu Enterprises, 2010.

[54] G.E. Lloyd, *Methods and Problems in Greek Science.* Cambridge University Press: Cambridge, 1999.

[55] Ptolemy, *Tetrabiblos.* Harvard University Press: Cambridge, Massachusetts, 1941.

[56] A. Jones, "Ptolemy's Mathematical Models and their Meaning." *In: Mathematics and the Historian's Craft.* Springer: Berlin, 2005, pp. 23-42.
[http://dx.doi.org/10.1007/0-387-28272-6_3]

[57] H.D. Rutkin, "The Use and Abuse of Ptolemy's Tetrabiblos in Renaissance and Early Modern Europe: Two Case Studies (Giovanni Pico della Mirandola and Filippo Fantoni)." *In: Ptolemy in Perspectives.* Springer: Netherlands, 2010, pp. 135-149.
[http://dx.doi.org/10.1007/978-90-481-2788-7_6]

[58] B.R. Goldstein, "Saving the phenomena: the background to Ptolemy's planetary theory", *J. Hist.*

Astron., vol. XXVIII, pp. 3-22, 1997.

[59] A. Einstein, *The World As I See It.* Covici Free Publishers: New York, 1934.

[60] A. Murschel, "The structure and function of Ptolemy's physical hypotheses of planetary motion", *J. Hist. Astron.,* vol. XXVI, pp. 33-61, 1985.

[61] C.M. Linton, *From Eudoxes to Einstein: A History of Mathematical Astronomy.* Cambridge University Press: Cambridge, 2010.

[62] G. Saliba, *Islamic Science and the Making of the European Renaissance.* MIT Press: Cambridge, Massachusetts, 2007.

[63] R.F. Jamil, "Copernicus and his Islamic predecessors", *Hist. Sci.,* vol. 45, pp. 65-81, 2007. [http://dx.doi.org/10.1177/007327530704500103]

[64] N. Copernicus, *On the Revolutions of the Heavenly Bodies.,* E. Rosen, Ed., Macmillan: London, 1972.

[65] P. Chaunu, *La Civilisation de l'Europe Classique.* Les Editions Arthaud: Paris, 1984.

[66] E.A. Burtt, *The Metaphysical Foundations of Modern Science.* Dover Publications: N.Y, 2003.

[67] G. Galilei, *Dialogue Concerning the Two Chief Systems – Ptolemaic and Copernican.,* S. Drake, Ed., The Modern Library: New York, 2001.

[68] M.L. Lial, C.D. Miller, and J.E. Hornsby, *Beginning Algebra.* Harper Collins College Publishers: New York, 1995.

[69] Plato, *Timaeus.,* D.J. Zeyl, Ed., Hackett Publishing Co: Indianapolis and Cambridge, Massachusetts, 2000. [386 BC]

[70] E. Husserl, *The Crisis of European Sciences and Transcendental Phenomenology.,* D. Carr, Ed., Northwestern University Press, 1970.

[71] E. Mach, *The Science of Mechanics: A Critical and Historical Account of its Developments.* Lightning Source Inc.: New York, 2007.

[72] J. Kepler, *Mysterium Cosmographicum (translated by F.M. Duncan as "The Secret of the Universe").* Abaris: New York, 1981.

[73] P. Barker, and B.R. Goldstein, "Theological Foundations of Kepler's Astronomy", *Osiris,* vol. 16, no. 1, pp. 88-113, 2001. [http://dx.doi.org/10.1086/649340]

[74] Yu.A. Danilov, and J.A. Smorodinsky, "Johannes Kepler: from "Misterium" to "Harmony"", *Soviet Science: Uspekhi,* vol. 109, no. 1, pp. 175-209, 1973. [in Russian].

[75] J. Kepler, "Astronomia Nova." *In: Selections from Kepler's Astronomia Nova: A Science Classics Module for Humanities Studies.,* W.H. Donahue, Ed., Green Lion Press: Santa Fe, NM, 2005, pp. 1-96.

[76] I. Newton, *The Mathematical Principles of Natural Philosophy (translated into English by Andrew Motte. With preface of Mr. Roger Cotes).* Daniel Ades: New York, 1846.

CHAPTER 2

Einstein's 1905 'Annus Mirabilis': Unification of Statistical Mechanics, Maxwellian Electrodynamics and Classical Thermodynamics

Abstract: To figure out in what way Einstein's 1905 'annus mirabilis' (wonder year) writings hang together one has to take into account Einstein's strive for unity evinced in his vigorous efforts to unify maxwellian electrodynamics, statistical mechanics and classical thermodynamics. It is exposed that Einstein's ingenious approach to special theory of relativity creation, substantially distinguishing him from Lorentz's and Poincaré's impacts, turns out to be a mere milestone of implementation of maxwellian electrodynamics, statistical mechanics and thermodynamics unification programme. Since the ether conception was a substantial snag for Einstein's statistical thermodynamics design, the leading part in the programme was played by Einstein's 1905 audacious light quanta hypothesis. Though apparently influenced in his critical stand against the classical physics research tradition by David Hume and Ernst Mach, Einstein hold an overall eclectic philosophical position. Yet when it related to creative momenta, Einstein's actual 1905 standpoint was closer to Kantian epistemology. It is contended that the most important Kantian concept apt to comprehend the special theory of relativity creation and all Einstein's 1905 writings *as a whole*, as well as the *order* of their arrangement is Kant's regulative idea of the systematic Unity of Nature. Maxwellian unification of electricity, magnetism and optics as a presentiment of the Einsteinian one is comprehended.

Keywords: 1905, Ampére, Annus mirabilis, Classical thermodynamics, Einstein, Faraday, Helmholtz, Hertz, Hume, Light quanta, Lorentz, Mach, Maxwell, Maxwellian electrodynamics, Neo-Kantian epistemology, Poincaré, Regulative principles, Research programmes, Special relativity, Statistical mechanics, Unity of Nature, Weber.

1. INTRODUCTION: THE EINSTEINIAN REVOLUTION PRESENTIMENT: THE MAXWELLIAN UNIFICATION OF OPTICS, ELECTRICITY AND MAGNETISM

It is a common place that Maxwell's stupendous theory of electricity and magnetism was a stage of advancement of Faraday's original scientific research programme grounded on the field concept. The Maxwellian research programme

Rinat Nugayev
All rights reserved-© 2018 Bentham Science Publishers

had successfully provided prediction (due to James Maxwell) and verification (due to Heinrich Hertz) of the radio waves and have eventually surpassed the Ampére-Weber programme built upon the notorious and misleading concept of 'action at a distance'.

Yet a more meticulous and profound study of XIX-th century physics [1 - 3] enables one to challenge this cursory standpoint as a harsh oversimplification and to expound an amended version of it *via* the following telling arguments.

(i) *James Maxwell himself many times, starting from his first 'electric' 1856-58 paper and up to his 1873 masterpiece ("A Treatise on Electricity and Magnetism"), had stubbornly underscored that the pivotal ideas of the Ampére-Weber research programme were as fruitful and efficacious for the advancement of electrodynamics as those of the field theories. Likewise, in accordance with his thoroughgoing projects, Maxwell in the upshot had contrived a hybrid theory amalgamating the components of the Ampére –Weber theory with that of Faraday.*

James Clerk Maxwell, for the first time, quoted Wilhelm Weber in a letter to his close friend and tutor William Thomson, on May 15, 1855. Due to Thomson's persistent and shrewd advice, Maxwell had meticulously studied Weber's "*Elektrodynamische Maasbestimmungen*" and punctuated:

> "I have been examining his [Weber's] mode of connecting electrostatics with electrodynamics, induction *etc.*, I confess I like it not at first...but I suppose the rest of his view are founded on experiments which are *trustworthy* as well as *elaborate*" (quoted from [4], p.150; my italics).

It is no wonder that at the outset of his diligent research in the theory of electricity and magnetism, in May 1855, a Cambridge post-graduate student wrote a thrill letter to his beloved father, stressing the significance of tentative study of the accounts of skillful German masters:

> "I am working away at electricity again, and have been working my way into the views of heavy German writers. It takes a long time to reduce to order *all* the notions one gets from *these* men, but I hope to see my way through the subject, and arrive at something intelligible in the way of a theory" (quoted from [5],p.105; my italics).

As young Maxwell diligently furnished his draft of "*On Faraday's lines of force*", he confessed to Thomson:

> "I am trying to construct two theories, mathematically identical, in one of which the elementary conceptions shall be about fluid particles *attracting at a distance* while in the other nothing (mathematical) is considered but various states of polarization *tension* and existing at various parts of space" (quoted from [6], p.241; my italics).

–All that a 'college brat', a young post-graduate could initially shyly controvert to the Ampére-Weber pre-eminent achievements was that

> "it is a good thing to have *two ways of looking at a subject*, and to admit that there are two ways of looking at it. Besides, I do not think that we have any right at present to understand the action of electricity, and I hold that the chief merit of a temporary theory is, that it shall guide experiment, without impeding the progress of the true theory when it appears" ([7], p.208; my italics).

Nevertheless, not long afterwards, Maxwell was able to retort by digging out more substantial arguments in favor of creation of new theory of electricity and magnetism. He shrewdly complained that the theory of Ampére and Weber is excessively mathematized and hence inevitably coarsens the profound and subtle links between the electric and magnetic phenomena. Likewise, it blindly oversimplifies *the true subtle relations between static and dynamic electricities*:

> "...the theory of the conduction of galvanism and that of the mutual attraction of conductors have been reduced to mathematical formulae, *but have not fallen into relation with the other parts of the science"* (Maxwell [7], p.155; my italics).

Further, recounting the genesis of his electromagnetic theory, in his fascinating introduction to "*A Dynamical Theory of Electromagnetic Field*" Maxwell deliberately renders an outline of theory of action at a distance:

This theory, as developed by MM. W. Weber and C. Neumann, is *exceedingly ingenious*, and *wonderfully comprehensive* in its application to the phenomena of statical electricity, electromagnetic attractions, induction of currents and diamagnetic phenomena; and it comes to us with the more authority, as it has served to guide the speculations of one who has made so great an advance in the practical part of electric science, both by introducing a consistent system of units in electrical measurement, and by actually determining electrical quantities with an accuracy hitherto unknown.

The mechanical difficulties, however, which are involved in the assumption of particles acting at a distance with forces which depend on their velocities are such as to prevent me from considering this theory as an *ultimate* one, though it may have been, and may yet be *useful* in leading to the *coordination* of phenomena" (Maxwell [8], p.527; my italics).

And eventually in his epoch-making "*Treatise*" Maxwell recounts the history of contrivance of his epoch-making system of equations as follows:

> "I was aware that there was supposed to be a difference between Faraday's way of conceiving phenomena and that of the mathematicians, so that *neither he nor they* were satisfied with each other's language. I had also the conviction that *this discrepancy did not arise from either party being wrong"* (Maxwell [9], p.499; my italics).

Likewise, in one of his last scientific writings – in 1879 review of George Fitzgerald's paper – Maxwell meekly portrayed his treatment of the Faraday effect in his ground-breaking 1861-62 paper as a '***hybrid***' one that hastily mingled his electromagnetic theory of light with queer components of an elastic solid theory. He took light waves as actual motions of the ether and elicited how these would disturb the spinning of the magnetic vortices in such a way as to prompt the plane of polarization of light to rotate. Accordingly, Maxwell regarded this detour into a "hybrid theory", amalgamating electrical and mechanical actions, the *least satisfactory* part of his comprehension of the Faraday effect ([10], p.18).

(ii) *No-one at Maxwell's Cavendish laboratory had made an attempt to corroborate the theory of their pre-eminent leader and benefactor [11]. Though Maxwell boldly and stubbornly conjectured that light constitutes a transverse electromagnetic wave, his whimsical hypothesis did not imply that he actually believed that light could be incited electromagnetically. He obstinately abstained from commenting on generation and detection of electromagnetic waves. Furthermore, there are grave reasons to uphold the view that Maxwell took the electrical production of the electromagnetic waves as an apparent impossibility ([10, 12]). And his renouncement was reinforced by the so-called "Maxwellians".*

It took almost a quarter of a century before Heinrich Hertz, a student and a partisan of Hermann Helmholtz, discovered radio waves. And up to 1888 Hertz did not even regard himself a true follower of Maxwell [3].

Note that in *all* his experimental papers Hertz systematically tried to circumvent quoting Maxwell. Likewise, in his conspicuous 1887 paper "*On Very Rapid Electric Oscillations*" Maxwell was not cited at all.

Furthermore, in Hertz's startling 1888 paper "*On the Electromagnetic Waves in Air and their Reflection*" Maxwell's theory of electromagnetism is quoted only once and after the following judicious reservation

> "I have described the present set of experiments, as also the first set on the propagation of induction, *without paying special regard toany particular theory*; and indeed, the demonstrative power of the experiments is independent of any particular theory" ([13],136; my italics).

And in Hertz's pre-eminent introduction to the first collected volume on "*Electrical Waves*" (1893) it is famously asserted that

> "Many a man has thrown himself with zeal into the study of Maxwell's work, and, even he has not stumbled upon unwanted mathematical difficulties, has nevertheless been compelled to abandon the hope of forming for himself an *altogether consistent* conception of Maxwell's ideas. I have fared no better myself" ([14], p.20; my italics).

The last quotation can be duly comprehended in the light of the fact that Hertz had meticulously planned his experiments in 1886-1887 for testing his teacher's ideas, *i.e.* the theory of Helmholtz, – but not Maxwell's one (see [3, 15] for detailed comments). The theory of Helmholtz was similar to that of Maxwell in that it was also a hybrid one skillfully blending 'field' notions with the notions of theory of 'action at a distance'.

The same hallmark – the "**hybrid**" nature of Maxwell's theory – was more than once punctuated by such luminaries as Henri Poincaré and Ludwig Boltzmann (see Heinrich Hertz's comments in [14], p.26).

Moreover, in the course of the Helmholtz programme rapid advancement, Helmholtz's star pupil Heinrich Hertz had skillfully derived Maxwell's equations from the modified version of theory of action at a distance ([16], pp.288-289).

(iii) *Hertz's startling 1887-1888 experiments cannot be taken as "crucial" ones prompting the definite choice between Ampére-Weber and Maxwell's rival research programmes.*

As Heinrich Hertz himself had complained,

> "Notwithstanding the greatest admiration for Maxwell's mathematical conceptions, I have not always felt quite certain of having grasped the physical significance of his statements. Hence it was not possible for me to be guided in my experiments directly by Maxwell's book. *I have rather been guided by Helmholtz's work*, as indeed may plainly be seen from the manner in which the experiments are set forth"([14], p.20; my italics).

(iv) *Michael Faraday's indubitable influence on James Maxwell was strongly exaggerated(probably for the most decent –patriotic- reasons*).

The comprehension of the acceptance of the field concept due to total devotion to intermediate action is not reinforced by meticulous study of Maxwell's writings [17]. The study elucidates that Maxwell started to regard the field notion as a pivotal means of amalgamating optics and electromagnetism sufficiently late. It was only *after* he had successfully derived the existence of the electromagnetic waves from his equations, *i.e. after* the invention of the so-called 'displacement current'. Up to that point he took the field notion only as an illustrative means for constructing the pictorial images of complicated vector differential equations.

Moreover, one should take into account that Faraday's deft apparatus of 'lines of force' and queer 'strains in the field' seemed too cumbersome and obsolete to his advanced contemporaries, especially when compared with subtle, precise and elegant action-at-a-distance theories of German and French masters [10].

Nevertheless, all the above mentioned arguments do not mean that I accept *all* the important conclusions of Margaret Morrison, Daniel Siegel and Olivier Darrigol splendid studies [1 - 3]. In my humble opinion, the main weakness of their startling accounts is consecutive *underestimation of Maxwell's original methodology* worked out for his breakthrough project of electricity, magnetism and optics synthesis. However, it is proverbial that in *any* field of creativity (and even in metaphysics and epistemology) Maxwell obstinately took his own way. Consider, for instance, the following part of his Marichal college speech appealing to his students.

> "It is best that every man should be settled *in his own mind*, and not be led into other men's ways of thinking under the pretense of studying science" (quoted from [11], 70; my italics).

Besides, as the author of "*A Treatise on Electricity and Magnetism*" confessed to his friend Litchfield in one of his letters, "I find I get fonder of metaphysics and less of calculations continually and my metaphysics are fast settling into the rigid high style, that is about ten times as far above Whewell as Mill is below him or Comte or Macaulay below Mill" (quoted from [5], 298).

Likewise, in his acute review of one of his friend's book, Maxwell admits that

> "We are glad to find, however, that in spite of the contempt which Prof. Tait pours upon the a priori physics of non-experimental philosophers, he admits that *there is a true science of metaphysics* which discusses the fundamental ideas of all science and knowledge, not by shutting out all the facts of experience, but by calling in all the evidence obtainable from the whole circle of science" (quoted from [5], 309; my italics).

Note that Maxwell's library contained extensive and carefully studied volumes of Aristotle, Dugald Stewart, Adam Smith, William Hamilton, John Wilson and, of course, Immanuel Kant and William Whewell.

Finally, one can recall Gustav Kirchhoff's blistering comment: "He is a genius, but one has to check his calculations".

Maxwell's scientific comportment, his *modus operandi* exactly fit a dictum of his Trinity college guru William Whewell:

> "Physical discoverers have differed from barren speculators, not by having no metaphysics in their heads, but by having *good metaphysics* while their adversaries had bad; and by binding their metaphysics to their physics, instead of keeping the two asunder" ([18],p. X; my italics).

In my judgement, one should take Ludwig Boltzmann's comments on Maxwell's vague writings more seriously. In his superlative lectures on Maxwell's theory of electromagnetism, as well as in his keen and exact comments on Maxwell's electromagnetic writings, the luminary of physics and one of the fathers of statistical mechanics complained that too many of Maxwell's writings, but especially his early electrical papers "*were not properly comprehended*" [19]. Maxwell's original synthetic methodology that arose from an intention to follow his Edinburgh philosophical teacher, Sir William Hamilton, in seeking efficacious compromise between the extremes of Kantian robust relativism and Scottish sustained "common sense realism" was an integral part of his pre-eminent itinerary of unification of optics and electromagnetism.

Maxwell's original research programme did surpass that of Ampére-Weber because it was a "*synthetic*" one (in the sense that was in more detail disclosed in the first chapter). It appeared, according to one of Maxwell's Trinity College philosophical teachers (who was himself a Kantian), one of "successive steps by which we gradually ascend in our speculative views to a higher and higher point of generality" (Whewell [18], vol. 2, 74). Contrary to Maxwell's, the Ampére-Weber programme was a *reductionist* one (see the first chapter for details) for it stubbornly sought to reduce all the theoretical ontologies to one and the same ontology of "action at a distance". In this connection the next passage from Ludwig Boltzmann's '*Werke*' is especially apt:

> "It is certainly useful to set up Weber's theory as a warning example for all times that we should always preserve the necessary ***mental flexibility***".

In his thought-provoking writings Boltzmann constantly stressed the need for a "*plurality of approaches*", including both mathematical formalism and picture-based physical theories. In particular, Maxwell's programme was not only efficacious to absorb the premises of the Ampére-Weber hard core, amalgamating them with Faraday's 'field' notions, as well as with those of Fresnel-Young optics; it was open for fusions with other research traditions as well. As Heinrich Hertz had put it,

> "From the outset Maxwell's theory excelled all others in elegance and in *the abundance of the relations* between the various phenomena which it included. The probability of this theory and therefore the number of its adherents increased from year to year" (Hertz [16], 19; my italics).

This quasi-Feyerabendian "*abundance of the relations*" was owing to that Maxwell did put forward as a synthetic principle the idea, that differed *radically* from that of Ampére-Weber by its flexible and contra-ontological, strictly epistemological, genuinely Kantian character. By referring everything to the purely *geometrical* idea of the motion of an imaginary fluid, I hope to attain *generality* and precision, and to avoid the dangers arising from a premature theory professing to explain *the cause* of the phenomena" (Maxwell [7], 159; my italics).

Evidently, Maxwell was not the first at aiming at optics and electromagnetism unification. Yet he renounced the ways his eminent predecessors had amalgamated electricity and magnetism. Why? What were the sober reasons?

The following eloquent quotation can help to find a judicious answer: the theories of action at a distance were too *abstract* to grasp the profound and intricate links between the electromagnetic phenomena.

> "The present state of electrical science seems *peculiarly unfavorable* to speculation...No electrical theory can now be put forth, unless it shows the connexion not only between electricity at rest and current electricity, but between the attractions and inductive effects of electricity in both states [...] The results of this simplification may take the form of a purely mathematical formula or of a physical hypothesis. In the first case we entirely lose sight of the phenomena to be explained; and though we may trace out the consequences of given laws, we can never obtain *more extended views of the connexions of the subject"* (Maxwell [7], p.155; my italics).

The circle of Maxwell's stupendous forerunners embraces Oersted, Ampére, Weber, Faraday and Thomson (lord Rayleigh). Yet James Clerk Maxwell's Weltanschauung was first and foremost distinguished by the *unsurpassed level of philosophicalprofundity*. A prodigy at the University of Edinburgh and a brilliant post-graduate at Trinity, Cambridge, lord Maxwell was enchanted by a profound and consistent skepticism of David Hume, George Berkeley and especially Immanuel Kant that he greedily absorbed at the lectures of Sir William Hamilton on 'mental philosophy' at the University of Edinburgh.

Hamilton's thought-provoking and splendid lectures "interested him greatly". Sir Hamilton's personal charisma and broad erudition made a lasting impression on him, in inciting "the love of speculation to which his mind was prone" [5].

Hamilton's opus magnus was *"Philosophy of the Unconditioned";* Sir William's philosophical system surely had a strong Kantian bias. Like Kant, he maintained that one can have a reliable knowledge only of "the relative manifestations of an existence, which in itself is our highest wisdom to recognize as beyond the reach of philosophy" ([20], p. 360). Yet on the other hand, unlike Kant, he had contended for the domestic standpoint of a 'natural realism' in the Reidian wake and his ingenious philosophical system can be evaluated as a "strange compound" [21], a whimsical blend of Reid and Kant, of domestic 'traditionalism' and European 'modernism'.

The glaring relativism or 'phenomenalism' that Hamilton derived from Kant and obstinately sought to engraft upon Scottish domestic philosophy was evidently absent from the original Scottish doctrine. Thus, renouncing Hume's fierce skepticism, Hamilton strenuously attempted to *find a reasonable compromise* between Kant's subtle relativism and Reid's oblique 'common sense' realism. It is this author's view that it was namely that peculiarity Maxwell had in mind indicating a pivotal tenet of his metaphysical programme on moving from Edinburgh University to Cambridge one:

> "in the meantime I have my usual superfluity of plans...4. Metaphysics – *Kant's Kritik of Pure reason in German, read with a determination to make it agree with Sir W. Hamilton...*" ([5], p.77; my italics).

The 'Copernican revolution' in epistemology famously engendered by Immanuel Kant consisted as a matter of fact in that the world of common experience had once and for all lost its dominating position in comprehending things that can be perceived by our senses. Kantian 'constructivism' had boldly exchanged the world of every-day experience by phantom realm of Galilean experimental and mathematical physics grounded on blatant idealizations of the common experience. In Kantian epistemology truth turned out to be something not spontaneously revealing and disclosing itself but something that can be duly understood only by subtle and rigorous 'scientific method'. Correspondingly, one can grasp not 'the things by themselves' or '*noumena*', but only the 'phenomena'. Thus, it is necessary to renounce the opportunity of comprehending the marrowbones of the Absolute Truth. Our sensory data are by no means representations of 'things in themselves', yet merely of the ways in which they appear to us. Thus the so-called '*analogies of experience*' are of especial significance for Kantian epistemology.

> "This type of cognition is cognition according to *analogy*, which surely does not signify, as the word is usually taken, an imperfect similarity between two things, but rather a perfect similarity between two *relations* in wholly dissimilar things. Such is an analogy between the legal relation of human actions and the mechanical relation of moving forces: I can never do anything to another man without giving him a right to do the same to me under the same condition; just as a body cannot act on another body with its motive force without thereby causing the other body to react just as much on it. Right and motive forces are here completely dissimilar things, but in their relation is there nonetheless complete similarity. By means of such an *analogy* I can therefore provide a concept of a relation to things that are absolutely unknown to me" (Kant [22], pp.146-147; my italics).

In "*The Critique of Pure Reason*", Kant scrutinizes a more important example:

> "Prior to the perception, however, and therefore completely a priori, we are able to cognize its existence, provided it stands in connection with some perceptions according to the principles of the empirical conjunction of these, that is, in conformity with the analogies of perception. For, in this case, the existence of the supposed things is connected with our perception in a possible experience, and we are able, with the guidance of these analogies, to reason in the series of possible perceptions from a thing which we do really perceive to the thing we do not perceive. Thus, we cognize the existence of a *magnetic matter* penetrating all bodies from the perception of the attraction of the steel-fillings by the magnet, although the constitution of our organs renders an immediate perception of this matter impossible for us" (Kant [23], p.170; my italics).

It should be stressed that it is *this striking example of the analogy of experience that was drawn by Kant from the purview of electromagnetism* thus paving the way to Maxwell's outrageous efforts. The author of "*A Treatise on Electricity and Magnetism*" had punctuated more than once that things one can measure directly, like mechanical forces, are only the outward manifestations of more profound processes, embracing entities like strengths of electric and magnetic fields, which are far *beyond* our common power of visualization [11].

A more detailed outline of Maxwell's original scientific research programme that he had pursued through all his life is given in his startling epistemological writings and first and foremost in his profound lecture "*Are There Real Analogies in Nature?*" read at the 'Apostles' elitist Cambridge club in 1856 (not long after the publication of his path-breaking paper "*On Faraday's Lines of Force*", 1855-56) – and in his epistemological chef-d-euvre '*Helmholtz*' [24] first published in 1877.

The Cambridge splendid lecture is not a succinct recap of Kantian epistemology but as a matter of fact a tense Maxwell's contention with 'Kant in himself.' It is no wonder that the very heading of the lecture turns out to be a question and not an assertion:"*Are There Real Analogies in Nature*?" Are they?

Maxwell gives no definite and unambiguous answer - in full accordance with Kant's pre-eminent antinomies that occur to Human Reason as vain attempts to trespass the clear-cut Limits of Experience. He multiplies subtle arguments pro et contra the assertion that certainly there *are* real analogies in Nature [5].

Maxwell's profound philosophical reflections in categories of Kantian antinomies are not accidental. In the wake of Sir Hamilton's tradition, Maxwell tries to conceive *his own epistemological way* between the Scylla of Kantian refined apriorism, relativism and transcendentalism and the Charybdis of Scottish domestic common sense realism.

Nowadays the genuinely Scottish view of knowledge is recapitulated in the following interconnected and subtle tenets [25, 26].

1. All human knowledge is necessarily *relational.*
2. *Analogies* are among the dominating relational ways of knowing.
3. Analogies are useful first and foremost for psychological reasons. For most men of science, understanding implies the implementation of analogies for *simplifying* and *organizing* human knowledge.
4. Vigorous psychological trends in the Scottish Common Sense tradition encourage *reconciliation* with logical and analytical biases of Kantian epistemology.

Thus for Maxwell the epistemological resolution of the glaring chasm of Kantian antinomies consists in cautious adopting *partial points of view*. No ultimate 'absolute' truth is attainable. What remain is humble eliciting correspondences or analogies.

> "Whenever they see a relation between two things they know well, and think they see there must be a similar relation between things less known, they reason from the one to the other. This supposes that although pairs of things may differ widely from each other, the *relation* in the one pair may be the same as that in the other. Now, as in a scientific point of view the *relation* is the most important thing to know, a knowledge of the one thing leads us a long way toward a knowledge of the other. If all that we know is relation, and if all the relations of one pair of things correspond to those of another pair, it will be difficult to distinguish the one pair from the other, although not presenting a single point of resemblance, unless we have some difference of relation to something else whereby to distinguish them. Such mistakes can hardly occur except in mathematical and physical *analogies"* (Maxwell; quoted from [5], p.124; my italics).

Eventually, Maxwell's original methodology of theory contrivance can be reconstructed as residing in the following profound principles.

(I) "***The Principle of Relational Nature of Truth***": It asserts that the *relation* is the most substantial thing to learn. It should be pointed out that even the examples of the most important analogies are *directly* drawn by Maxwell from Kant's "*Prolegomena"*.

(II) "***The Principle of Theory Laidenness of Observation***": It is deftly extracted from Immanuel Kant's "*Critique of Pure Reason"*, "*Prolegomena"* and William Whewell's *"Philosophy of Inductive Sciences"* and maintains that

> "The dimmed outlines of phenomenal things all merge into one another unless we put on the *focusing glass of theory*, and screw it up sometimes to one pitch of definition and sometimes to another, so as to see down into different depths through the great millstone of the world" (Maxwell; quoted from [5], 125; my italics).

Thus, all the phenomena of Nature are interconnected and merge into one another. Hence all the contentions between various theoretical accounts are due to the fact that they focus on the different aspects and different levels of the phenomena under investigation. Hence a theoretician's aim is to elucidate the '*appropriate ideas*' (William Whewell's term!) to cover the various domains of experience. But where should one excavate them? – In experience, as springing out of direct generalizations of the experimental data? –Another eloquent passage from

Maxwell – a part of his 1854 letter – is especially relevant here:

> "It is hard work grinding out 'appropriate ideas', as Whewell calls them. I think they are coming out at last, and by dint of knocking them against all the facts and *half-digested theories* afloat, I hope to bring them to shape, after which I hope to understand something more about *inductive philosophy* that I do at present."

> "I have a project of sifting the theory of light and making everything stand upon definite experiments and definite assumptions, so that things may not be supposed to be assumptions when they are either definitions or experiments" (Maxwell [5],112; my italics).

Maxwell elicits that the 'appropriate ideas' are not the slavish copies of things; rather they constitute the *a priori* forms by which a ravel of sensations is diligently "brought to order".

Initially the 'appropriate ideas' are really 'vague and dim' (due to their commonsense origin); yet in the long run they are gradually 'grinded out' by knocking them with the 'facts' and with the other theories that stand behind them. Nevertheless, the theoretician's aim is not only to contrive and polish the subtle notions referring to the different aspects of the natural phenomena, but also to *reconcile* the notions in comprehensive *synthesis*.

The make-up and the milestones of such a unification are meticulously disclosed in Maxwell's thought-provoking paper "*Hermann Ludwig Ferdinand Helmholtz*" that starts as follows:

> "Hence the ordinary growth of human knowledge is by accumulation round a number of *distinct centers*. The time, however, must sooner or later arrive when two or more departments of knowledge can no longer remain independent of each other, but must be *fused* into a ***consistent whole***. But though men of science may be profoundly convinced of the necessity of such a fusion, the operation itself is a most arduous one. For though the phenomena of nature are all consistent with each other, we have to deal not only with these, but with the hypotheses which have been invented to systematize them; and it by no means follows that because one set of observers have

> labored with all sincerity to reduce to order one group of phenomena, the hypotheses which they have formed will be consistent with those by which a second set of observers have explained a different set of phenomena. Each science may appear tolerably consistent within itself, but before they can be combined into one, each must be *stripped of the daubing of untempered mortar* by which its parts have been prematurely made to cohere." (Maxwell [24], 592; my italics).

(III) In other writings Maxwell points out the **significance** of the next basic tenet - "***The Principle of Cross-Fertilizationof the Sciences***" evoking the image of bees assiduously pollinating flowers [27].

Note that some important premises of Maxwell's methodology were prompted by apparent influence of William Whewell (1794-1866), a polymath master of Trinity College, Cambridge. The list of Whewell's profound writings includes his opus magnus "*Philosophy of the Inductive Sciences*" (first published in 1840) and "*Novum Organon Renovatum*" (1858-1869).

Whewell's epistemology is grounded on the assertion that all Human Knowledge has both an ideal, or *subjective* dimension, and an *objective* one. Correspondingly, the 'fundamental antithesis' of knowledge consists in that "in every act of knowledge…there are two opposite elements, which we may call Ideas and Perceptions" ([18], 307). Though Whewell's overall philosophical standpoint is commonly characterized as a breed of Kantianism, Whewell bluntly criticized Kant and German Idealists for their exclusive focus on the Ideal (subjective) component of knowledge. On the other hand, he strongly criticized John Locke and the partisans of empiricist epistemological tradition for their exclusive focus on the Perceptive (objective) element. Following Francis Bacon, he claimed to be seeking a 'third way' between the extremities of pure Rationalism and Lock's ultra-Empiricism. In my view, it is this propensity to avoid the extremities that influenced Maxwell's own philosophical standpoint.

According to Whewell, each science has a particular Fundamental Idea which is necessary to organize the facts which that science is concerned. For instance, Space is the Fundamental Idea of geometry, Cause is the Fundamental Idea of mechanics, and so on. Any Fundamental Idea has certain concrete 'conceptions' included within it; the latter are 'special modifications' of the Fundamental Idea applied to particular types of circumstances. For instance, the conception of force is a subtle modification of the key Fundamental Idea of Cause, *etc*. The advancement of a mature theory eventually consists in applying the Fundamental Idea to the description of each new domain of empirical reality.

It is important for a Kantian that Fundamental Ideas are "not a consequence of experience, but a result of the particular constitution and activity of the mind, which is independent of all experience in its origin, though constantly combined with experience in its exercise" (Whewell [18],vol.1,91).

Now any observation is 'idea-laden' since all observations necessarily involve 'unconscious inferences' involving the Fundamental Ideas ([18], vol.1, 46). In that respect Whewell's Fundamental Ideas are quite similar to Immanuel Kant's 'forms of intuition' and correspondingly to James Maxwell's original methodology. Just to quote the author of "*Philosophy of Inductive Sciences*":

> "The fact is merely the occasion by which the engine of discovery is brought into play sooner or later. It is, as I have elsewhere said, only the spark which discharges a ***gun*** already loaded and pointed" ([18], vol.2, 23; my bold italics).

So, now we can understand where Maxwell's famous expression (from a letter to his cousin C.H. Cay) "great guns" comes from:

> "I also have a paper afloat, with an electromagnetic theory of light, which, till I am convinced to the contrary, I hold to be ***great guns***" (quoted from [11],123; my bold italics).

Yet the most fascinating part of Whewell's original doctrine is his comprehension of the 'induction'. To begin with,

> "By the sense of sight we see various shades and colors and shapes before us, but the *outlines* by which they are separated into *distinct* objects of definite forms, are the work of the mind itself... When we see a body move, we see it move in a path or *orbit*, but this orbit is not itself seen; it is constructed by the mind. In like manner when we see the motions of a needle towards a magnet, we do not *see* the attraction or force which produces the effects; but we infer the force, by having in our minds the Idea of Cause"([18],vol.1,26).

Hence for Whewell in induction "there is a New Element added to the combination [of instances] by the very act of thought by which they were combined" ([18], vol.2, 48). This new element is '***colligation***'. The latter constitutes the mental operation of bringing together a number of empirical facts by 'superinducing' upon them a conception which unites the facts and renders them capable of being expressed by a general law.

> "Thus in each inference made by Induction, there is introduced some General Conception, which is given, not by the phenomena, but by the mind...In order to obtain our inference, we travel beyond the cases which we have before us; we consider them as mere exemplifications of some Ideal Case in which the relations are complete and intelligible. We take a Standard, and measure the facts by it; and this Standard is *constructed* by us, not offered by Nature" ([18], vol.2, 49).

Typically, finding the appropriate conception with which to colligate a class of phenomena requires a whole series of inferences, a "train of researches" ([18], vol.1, 297).For Whewell, types of inference in colligation are: (1) enumerative; (2) eliminative; (3) analogical.

Furthermore, once a theory is invented by means of 'colligated induction', it must pass a whole multitude of tests before it can be regarded as an empirical truth. These tests are (i) prediction; (ii) consilience of inductions; (iii) coherence.

While the significance of the (i) point is obvious, the (ii) one is elicited as follows:

> "That rules springing from remote and unconnected quarters should thus leap to the same point, can only arise from that being the point where truth resides" ([18],65).

In my view, Maxwell's startling discovery that the velocity of electromagnetic waves is equal to the speed of light, is an excellent example of Whewell's 'consilience of inductions'. In general, Whewell's influence on Maxwell's methodology is caused by their common source – Kantian epistemology.

Note that "*A Treatise on Electricity and Magnetism*" (1873) was simultaneously an encyclopedia and a textbook. Rather the most substantial results relating to electromagnetism were derived first in three epoch-making papers: "*On Faraday's Lines of Force*" (1856-58), "*On Physical Lines of Force*" (1861-62) and "*A Dynamical Theory of Electromagnetic Field*" (1864).

The first paper meticulously worked out the "*analogies*" method grounded in Kantian epistemology. The method decidedly renounces the 'ontological' accounts looking for the 'marrowbones' of phenomena of electricity and magnetism and asserting that electricity and magnetism in reality are 'fields' and not 'action at a distance' entities, or *vice versa*. The thrust of Maxwell's approach is to take Faraday's lines of force as a kind of queer *imaginary* tubes filled with an ***ideal*** incompressible fluid.

"I propose then, [...]; and lastly to show how by an extension of these methods, and the introduction of another idea due to Faraday, the laws of the attractions and inductive actions of magnets and currents may be clearly conceived, *without making assumptions as to the physical nature of electricity*, or adding anything to that which has been already proved by experiment.

By referring everything to the purely geometrical idea of the motion of an imaginary fluid, I hope to attain generality and precision, and to avoid the dangers arising from a premature theory professing *to explain the cause of the phenomena*" (Maxwell [7], 159; my italics).

It is substantial for a true Kantian that the incompressible fluid can have nothing to do with experimental reality. The necessary constraint on the theory consists in that it should be *coherent*, that is the mathematical constructs (or theoretical abstract objects) should not contradict each other. In all the other respects the method of physical analogies allows such an unlimited freedom of imagination that even the conservation laws can be freely broken down.

> "There is nothing *self-contradictory* in the conception of these sources where the fluid is created, and sinks where it is annihilated. The properties of the fluid are *at our disposal*, we have made it incompressible, and now we suppose it *produced from nothing* at certain points and reduced to nothing at others" (Maxwell [7], 162; my italics).

Maxwell fosters the generality of the lines of force approach, for it can account for any kind of force. For instance, it does not renounce the force of action at a distance.

In the remaining parts of the 1856-58 paper Maxwell discloses the subtle ways by which the idea of incompressible fluid motion can be implemented in the domains of statical electricity, permanent magnetism, magnetism of induction, and uniform galvanic currents. The pivotal component of his innovations consisted in the invention of "*neutral language game*" for comprehension and comparison of the consequences from the competing theories. Evidently Maxwell's "neutral language" was not Carnap and Reichenbach's notorious 'observation language' provoked by direct generalizations of the "protokolsatze". Rather Maxwell is fully aware of the so-called 'theory – laidenness' of the observations at hand. As he puts it, "experimental laws already established, which have generally been expressed in the language of other hypotheses" ([28], 162). He easily admits that any observation necessarily carries the footprints of the theoretical language rendered to describe it. ("The daubing of untempered mortar", as he will aptly denote them in his "*Helmholtz*" masterpiece).

In order to embrace in a coherent theoretical scheme *all* the results of the different experiments carrying the footprints of 'incommensurable' theoretical languages, one should contrive an *artificial* theoretical language distant from the subtle languages of theories under comparison. In Maxwell's case this language turned out to be the language of solid state mechanics (including hydrodynamics). Maxwell's ultimate aim was to rewrite all the known (empirical and theoretical) laws of electricity and magnetism on the basis of the neutral language of hydrodynamics and then to compare them in order to contrive a coherent theoretical system.

The final result of the 1856-58 paper turned out to be a system of equations without the '*displacement current*'. One should not be surprised to learn that one of the main weaknesses of the theory of incompressible fluid consisted in that the latter, apart from some peculiar cases, was unable to comprehend the interactions of electrical and magnetic fields and electric currents. Likewise, it could not explain Faraday's 1845 startling link between the phenomena of optics and electromagnetism.

The ultimate goal of Maxwell's original research programme was to disclose the crucial link 'between electricity at rest and current electricity' that was absent in the Ampére-Weber electrodynamics. Was this ambitious aim attained in 1856-58 paper? – Certainly not. The important link between the current density $\mathbf{j}$ and the charge density ρ was absent in Maxwell's primordial 1856 theoretical scheme. It was to appear later, after the invention of the 'displacement current' and disclosing its consequence – the continuity equation $\text{div}\ \mathbf{j} + \partial\rho/\partial t = 0$.

In 1861 Maxwell started publishing his second paper in four fascinating parts. Its ultimate goal was to rederive the results of theories of Weber and Neumann drawing upon a new bold hypothetical model describing the *vortices of incompressible fluid* (Maxwell [28], 162).

More than once Maxwell had to warn the mistrustful reader that

> "the author of this method of representation does not attempt to *explain the origin* of the observed forces by the effects due to these strains in the elastic solid, but makes use of the mathematical analogies of the two problems to assist the imagination in the study of both" (Maxwell [28],163; my italics).

The theory was entrenched in W. Thomson's eminent investigations. The latter had famously excavated that the crucial link between the phenomena of electricity and magnetism strikingly resembles the mathematical form between certain parts of the phenomena, of which one has a *linear* and the other a *rotatory* character. Note that William Thomson contrived the theory of vortices in incompressible fluid aiming to provide a compelling explanation of Michael Faraday's startling experiments on the rotation of the plane of polarized light when transmitted along the lines of magnetic force. Thus, they were the strenuous efforts to comprehend and theoretically reproduce 'the Faraday effect' that provoked the inevitable *encounter* of the theory of optics and the theory of magnetism. In Maxwell's second theoretical model M_2 the magnetic field was represented now by a set of vortices in incompressible fluid with the axes of rotation coinciding with the direction of magnetic field at a point. Neutral language was implemented to display not tube hydrodynamics but a theory of stresses in the medium where the necessary relations among the forces are depicted by mathematical entities that are called *tensors* nowadays. The tensor that describes a stress commonly consists of a mixture of three principal pressures or tensions, in direction at right angles to each other. The tensor apparatus of solid state mechanics engendered a new neutral language game, that enabled to calculate the force acting upon an element of the medium:

$$\mathbf{F} = \mathbf{F}_1 + \mathbf{F}_2 + \mathbf{F}_3 + \mathbf{F}_4 + \mathbf{F}_5. \qquad \textbf{(i)}$$

The first term $\mathbf{F}_1$ describes the force acting on magnetic poles; the second term $\mathbf{F}_2$ refers to the action on bodies capable of magnetism by induction; the third $\mathbf{F}_3$ and fourth $\mathbf{F}_4$ terms jointly describe the force acting on electric currents; the fifth term $\mathbf{F}_5$ refers to the effect of simple pressure that lacks a direct electromagnetic analogy. It is this peculiarity that discloses with especial clarity the' toy' character of Maxwell's splendid model.

One of the most intricate problems of the vortices model M_2 consisted in the following: how the rotation from one vortex to another could be transmitted so that "vortices in a medium exist side by side, revolving in the same direction about parallel axes" [29]? – The only model at hand that helped Maxwell in describing this kind of motion was that of the vortices being separated by a layer of particles called the '*idle wheels*'. Was it possible to link these queer particles to electricity?

In the second part of the 1861-62 paper entitled "*The Theory of Molecular Vortices applied to Electric Currents*" the author arrives at the **toughest problem** of his account: what is "the physical connexion of these vortices with electric currents, while we are still in doubt as to the nature of electricity"? It is at this point where he has to admit the principal *limits* of pure mechanical account and to insert **the components of hostile action at a distance theory**! Or, implementing the methodological notions of [30], one can elucidate that Maxwell had to construct the '**crossbred theoretical objects**' from the languages of both theories, amalgamating the hallmarks of quite different theoretical schemes and rather 'incommensurable' (Kuhn, Feyerabend) research traditions.

In Maxwell's subtle theoretical model M_2, an electric current is displayed by the flow of the moveable particles interposed between the neighboring vortices. As a result,

> "These particles, in our theory, *play the part of electricity.* Their motion of translation constitute an electric current, their rotation serves to transmit the motion of the vortices from one part of the field to another, and the tangential pressures thus called into play constitute electromotive force. The conception of a particle having its motion connected with that of a vortex by perfect rolling contact may appear somewhat awkward. ***I do not bring it forward as a mode of connexion existing in nature***" (Maxwell [28], 345; my bold italics).

On constructing the abstract theoretical objects 'electrical particles' and 'electric current representing the motion of such particles' Maxwell had significantly demolished the spirit of Faraday's fine programme. In Faraday's *genuine* research programme, the electrical charges should be comprehended as created by the ends of lines of force; hence they lack an independent substantial existence. Correspondingly, in Faraday's research programme the electric current has to be taken not as motion of real particles but merely as an "energy axis". This is the marrowbone of the British field programme: the fields are primary, and the particles are only secondary. But Maxwell's eclecticism was perfected later on by H.A. Lorentz's refined dualism. Lorentz famously prompted it in a seminal 1875 paper: "I shall start with *instantaneous action at a distance*: thus we will be able to found the theory on the most direct interpretation of observed facts" (quoted in [3], 323; my italics). So it was not a temporary retreat at all. And after 1861 Maxwell deftly and stubbornly inserted the components of the Ampére-Weber research programme into his theoretical models many times [3].

Yet his shrewd advances were apparently insufficient since the derivation of Coulomb's law was apparently lacking. This stumbling block was successfully eliminated in the third part of 1861-62 paper entitled "*The Theory of Molecular Vortices applied to Statical Electricity*". Note that the vortices M_2 model contained too many ad hoc assumptions, and only some of them had been removed with a help of the so-called "*Maxwell's miracle*". Indeed, it turned out that if one transfers, in the course of encounter of the Fresnél optics and the theory of electromagnetism, the ether hallmarks from optics to electromagnetism, at least one blatant ad hoc premise can be successfully removed:

> "it is necessary to suppose, in order to account for the transmission of rotation from the exterior to the interior parts of each cell, that the substance in the cells possesses elasticity of figure, similar in kind, though different in degree, to that of observed in solid bodies. The undulatory theory of light requires us to admit this kind of elasticity in the luminiferous medium, in order to account for transverse vibrations. We need not then be surprised if the magneto-electric medium possesses *the same properties*" (Maxwell [28], 13; my italics).

As a result, the successful extrapolation of the theory of molecular vortices on the electrostatic domain became possible due to the elasticity of the vortices; it enabled the medium to maintain the elasticity waves. As a result,

> "The velocity of transverse undulations in our hypothetical medium, calculated from the electromagnetic experiments of M.M. Kohlrausch and W. Weber, *agrees so exactly* with the velocity of light calculated from the optical experiments of M. Fizeau, that we can scarcely avoid the inference that light consists of the same medium which is the cause of electric and magnetic phenomena" (Maxwell [28], 22; my italics).

The insertion of the displacement current was triggered by Maxwell's strenuous efforts to link the equations relating to electrical current to those of electrostatics.

It demanded the modification of the Ampére law for contrivance of a new one able to describe the elasticity of the vortices' medium. Maxwell's persistent efforts to embrace all the main empirical laws relating not only to electricity and magnetism but to optics as well were the impetus behind this change.

As a result Maxwell arrived at his famous system of equations along with the continuity equation exhibiting that electrical particles (that transform the rotations from one vortex to another) do not appear from nothing and cannot disappear to nowhere. However, there was no ultimate synthesis of optics and electromagnetism in 1861-62. One can fix only the *commencement of their reconciliation*, the outset of 'grinding' of fairly different theoretical ontologies.

And at last in 1864 Maxwell advanced a modified version of his 1861-62 theoretical model that tentatively avoided any special premises on the 'nature' of molecular vortices. He started to derive his famous equations from abstract dynamics of Lagrange. The Lagrangian function L is commonly taken as a difference between the kinetic and potential energies of a system. From the function L Maxwell was able to derive the basic wave equation of electromagnetism without any dubious assumptions about molecular vortices and forces between electrical particles. Although displacement retained a prominent position in "*A Dynamical Theory of Electromagnetic Field*", its role differed significantly from the role it played in the 1861-62 theory. It was no longer associated with the changes in positions of rolling particles. On the contrary, Maxwell took it simply as the motion of electricity, in lucid terms of a quantity of charge crossing a certain area.

Nevertheless, despite Maxwell's ambitious claim to provide deductions from (three) experimental facts, his account still required the *postulation* of a displacement current, something that could neither be verified by nor deduced from experience [2, 3].

Later Maxwell recapitulated the advances of 1864 paper in the letter to C. Hockin, September 7th, 1864 as follows:

> "I have also cleared the electromagnetic theory of light from all *unwarrantable assumptions*, so that we may determine the velocity of light by measuring the attraction between bodies kept at a given difference of potential, the value of which is known in electromagnetic measure" ([5], 168; my italics).

In the upshot, Maxwell's creative activity in electromagnetism ends with '*A Treatise on Electricity and Magnetism*' posited as an Encyclopedia of Electricity and Magnetism. In his "*Treatise"* Maxwell diligently proceeds in purifying his deductions from the remnants of his dubious ideal M_1 and M_2 models and in strengthening the abstract Lagrangian approach. In the final chapter XX of his '*Treatise*', that included the electromagnetic theory of light, he stressed the pivotal argument in defense of existence of electromagnetic waves as follows:

> "To fill all space with a new medium whenever any new phenomena is to be examined is by *no means philosophical*, but if the study of two different branches of science has independently suggested the idea of a medium, and if the properties which must be attributed to the medium in order to account for electromagnetic phenomena are of the same kind as those we attribute to the luminiferous medium in order to account for the phenomena of light, the evidence for the physical existence of the medium will be considerably strengthened.[...]

If it should be found that the velocity of propagation of electromagnetic disturbances is the same as the velocity of light, and this is not only in air, but in other transparent media, we shall have *strong reasons* for believing that light is an electromagnetic phenomenon..."([9], 781; my italics).

In his "*Treatise"* Maxwell was challenged by the same problem as in 1864 paper: the problem of implementation of Lagrangian mathematical formalism in the case of electromagnetic field. Nevertheless, a score of tense pages of his "*Treatise*" chapter hinge upon a detailed Lagrangian treatment only for the peculiar case of closed conduction currents that interact with each other. And when, two chapters later, he comes to build on his Lagrangian formulation to formulate the *general* equations of his electromagnetic theory, he *simply adds* the displacement to the conduction current 'by hands' to figure out the total current [12]. But as a matter of fact this shrewd trick undermined all the splendor of his Lagrangian method substantially. Note that the first *direct* experimental evidence in favor of existence of displacement currents emerged only with Hertz's experiments and culminated in the famous 1888 discovery of radio waves. As usual, the Lagrangian formulations were just retroactive attempts to accommodate the results obtained first by other rather dubious means.

Let us return to Maxwell's original synthetic research programme. Eventually Maxwell demonstrated that his queer elastic vortex medium *would* propagate waves whose velocity, calculated from electromagnetic constants, was that of light. Yet he was constantly silent on *how* electromagnetic waves might be generated, nor did he attempt to derive the complicated laws governing reflection and refraction [31]. Hence the hard task of extracting a cogent theory from the "*Treatise*" and of casting it into a form in which it could command general assent fell on the broad shoulders of 'the Maxwellians' [32].

Nevertheless, note that in a 1873 review of Fitzgerald's paper, Maxwell exposed his own treatment of the Faraday 1845 effect as a "*hybrid*" one in which he had blended his electromagnetic theory of light with the elastic solid theory. He had taken the light waves as actual motions of the ether and had investigated how these would disturb the spinning of the magnetic vortices in such a way as to cause the plane of polarization of the light to rotate.

In the above mentioned review Maxwell had characterized this detour into a "**hybrid theory**" (his own term!), in which electrical and mechanical actions were merged, as the *least satisfactory* part of his explanation of the Faraday effect. And Fitzgerald's startling 1879 paper had disclosed, more lucidly than before, the profound incompatibility between Maxwell's theory and an elastic ether.

> "The conclusion was inescapable: if Maxwell's theory were to survive, it had to be *cut loose* from reliance on an elastic solid ether and given a fundamentally new basis. Attempts to produce a 'hybrid' theory, such as Maxwell had pursued in his own account of the Faraday effect, had to be *abandoned*" ([10], 529; my italics).

Eventually, in his "*Treatise*", a pre-eminent Compendium of Electricity and Magnetism, Maxwell recapitulated his basic results. His Copernican deeds consisted in strengthening arguments for identification of electromagnetic and luminiferous ethers and invention of the **crossbred theory** with displacement current capable of unifying the theories of electromagnetism and optics.

Likewise, in his "*De Revolutionibus*" a Catholic canon Nicolas Copernicus had audaciously pioneered in taking the Earth as a common planet orbiting the Sun. Hence he had contrived a crossbred theoretical model capable of descending the mathematical principles from divine phenomena to the mundane ones. On the other hand, through the same crossbred object, the principles of Earth physics were genuinely ascended from mundane objects to the Skies [33].

In a similar vein, James Clerk Maxwell had contrived a queer crossbred object – the displacement current - and was able to extrapolate the electromagnetic principles on the optical phenomena, and *vice versa*. Setting force a kind of *"complementarity principle"* in chapter XXIII entitled *"Theories of Action at a Distance"*, Maxwell highlights the difference between field and corpuscular accounts as follows:

> "Now we are unable to conceive of propagation in time, except either as the flight of a *material substance* through space, or as the propagation of a condition of motion or *stress* in a medium already existing in space" (Maxwell [9], 488, my italics).

Thus, one is regrettably ignorant of what is really moving in the space between magnets and conductors, but if one decides to portray it, he has no other 'appropriate' images except the common stuff: pesky 'waves' and tiresome 'particles'. Maxwell's account apparently included the seeds of modern quantum theory. As Richard Feynman had maintained much later,

> "Well, it depends on our prejudices. Many physicists used to say that direct action with nothing in between was inconceivable. (How could they find an idea inconceivable when it had already been conceived?) [...] The only sensible question is what is the most *convenient* way to look at electrical effects. Some people *prefer* to represent them as the interaction at a distance of charges...Others *love* the field lines" ([34], 20; my italics).

And the lines of force still serve merely as a "crude way of describing field". They possess some indubitable merits since they provide a visual representation yet they have their own weaknesses too. For instance, when one talks about the lines of force of vectors **E** and **B,** one should not exaggerate the reality of their existence, since the lines may simply disappear when one tries to consider them in another reference frame.

2. EINSTEIN'S 1905 AND 1913-1915 EFFICACIOUS UNIFICATIONS

In Section 1 of the present chapter the arguments were strengthened in favor of the tenet according to which the genesis of maxwellian electrodynamics can be comprehended as a result of the encounter, reconciliation and intertwinement of the old pre-maxwellian research traditions'. The scientific research programmes to coordinate with each other had been the electrodynamics of Ampére-Weber, the wave theory of Young-Fresnél and Faraday's original programme. In particular, the crux of the section was that Maxwell's unification design could be successfully implemented since his programme had assimilated the ideas of the Ampére-Weber programme, as well as the premises of the Young- Fresnél and the Faraday research programmes. Eventually Maxwell's victory over his stupendous rivals became possible since the hard core of Maxwell's successful unification strategy had been permeated by the spirit of Kantian epistemology. In particular Maxwell had put forward as a pivotal synthetic principle the tenet that radically differed from that of the rival accounts by its open, flexible and contra-ontological, genuinely Kantian character.

Yet the search for a coherent theoretical basis for all of physics had not originated with Maxwell. His celebrated progenitor was Isaac Newton, and no less pre-eminent partisan was Albert Einstein. And, in my view, Newton's works, as well as the writings of Maxwell and Einstein, were all the necessary *momenta* of common '*de-ontologization*' process inherent in Modern Times. The process had commenced in XV-XVII centuries constituting the pivotal hallmark that distinguished the new science of Galileo and Newton from that of Aristotle and Ptolemy. It was namely the spirit of science of modernity that was later shrewdly comprehended by Immanuel Kant in his epoch-making "*Critique of Pure Reason*".

> "When Galileo rolled balls of a weight chosen by himself down an inclined plane, or when Torricelli made the air bear a weight he had previously thought to be equal to that of a known column of water, or when in a later time Stahl changed metals into calx and then changed the latter back into metal by first removing something and then putting it back again, a light dawned on all those who study nature. They comprehended that reason has insight only into what itself produces according to its *own design*; that it must take the lead with principles for its judgements according to constant laws and *compel nature to answer its questions*, rather than letting nature guide its movements by keeping reason, as it were, in leading strings" (Kant [23],108; my italics).

The sensible world, *i.e.* the world of appearances, is tentatively *constructed* (Konstruieren) by the Human Mind from subtle combination of recondite sensory matter that one receives passively and rigid and lucid *a priori* forms that are supplied by human cognitive faculties. All in all Kant's constructivist foundation for scientific knowledge restricts science to be the realm of appearances and maintains that a priori knowledge of 'things in themselves' is impossible.

Newtonian mechanics, maxwellian electrodynamics and Einstein's special and general theories of relativity were skillfully *fixed into* sweeping de-ontologization of physics grounded on the utmost rejection of Aristotelian visual ontology and the corresponding ways of theory building.

Aristotelian physics had been an empirical discipline par excellence that diligently accumulated common experience much better than the mathematical science of Galileo and Newton did [35]. Hence the decisive role in the genesis of modern science was played not so much by 'Experience' as by '*Experimentation*'. And the latter consists not so much in Thorough Observations of nature as in its Systematic Questioning. The questions should be carefully put on a peculiar language elucidated in Galileo's path-breaking "*The Assayer*" (1623):

> "Philosophy is written in that great book which ever lies before our eyes — I mean the universe — but we cannot understand it if we do not first learn the language and grasp the symbols, in which it is written. This book is written in the *mathematical language*, and the symbols are triangles, circles and other geometrical figures, without whose help it is impossible to comprehend a single word of it; without which one wanders in vain through a dark labyrinth" (Galilei [1623], quoted from [36], 75; my italics).

At the sake of mathematization Galileo radically transformed the methodology of natural science. He defiantly elevated 'idealization' and 'thought experiments' up to the ranks of the leading scientific methods [37]. In the long run it made possible for Galileo to formulate "the principle of inertia" and to come close to Newton's second law of dynamics (Mach [38]).

The very opportunity of implementation of mathematical methods in natural science turned out to be grounded on the procedure of idealization. Correspondingly scientists take all natural phenomena as more or less adequate approximations of some Platonic 'ideal essences'. The latter lack profound existence within the natural phenomena but can be freely constructed by human mind. And they are the relations between the ideal essences that are described by the Laws of Nature. At the same time the relations between real objects (*e.g.* rods and clocks) are exhibited by the approximations to strict laws of nature (Husserl [39]).

As Galileo succinctly put it, "the search for essences, in my judgement, is a vain and hopeless kind of pursuit". But if truth is acquired only in experience and we come to know not the 'things in themselves' but only the 'phenomena', one should flatly reject even the faintest hope of attaining the Absolute Truth.

According to the spirit of science of modernity, so keenly grasped by Kant, phenomena of nature do not constitute the complicated vehicles through which the essences of things show themselves up. On the contrary, natural phenomena constitute the 'essences in themselves' that contain *all* the possible information on the processes under study.

The next step in implementation of Galilean epistemological programme was taken by Newton who flatly rejected the search for the essences of gravitation phenomena. He had resolutely jettisoned the very opportunity of answering the question "*why* ponderable bodies attract each other?" and had famously provided instead a simple equation describing *how* ponderable bodies gravitate (see [40], 121 for details).

Then surely comes Maxwell himself who persistently abstained from revealing the essences of electrical and magnetic phenomena. In the path-breaking 1856-58 paper "*On Faraday's Lines of Force*" that coined the 'hard core' of his electromagnetic research programme he stressed that

> "the laws of the attractions and inductive actions of magnets and currents may be clearly conceived, without making assumptions as to the *physical nature* of electricity, or adding anything to that which has been already proved by experiment" (Maxwell [7], 159; my italics).

Since primary efforts to take the 'aether' as a medium for propagation of electromagnetic waves had failed, eventually Maxwell had to regard the ether as merely a component of the models intended for accumulation and classification of the corresponding 'facts' at best.

Yet another important function of ether was left inviolable – 'to be a carrier of absolute system of reference'. And it was Albert Einstein who had emancipated us of it. He exhibited that the conception of ether was a snag that hampered unitary comprehension of electricity and magnetism as different components of one and the same entity – the electromagnetic field tensor F_{ij}. It was possible since Einstein took the next step after Newton in the direction of refusal of contemplating the essences of space and time. And in 1915 he went even further reducing, for the sake of unification of gravity and inertia, the 'nature' of gravitational field to 'distortions of space and time'. The gravitational field strengths and the Coriolis inertia forces began to be comprehended as different components of a single curvature tensor R_{ij}. As Michele Janssen punctuates,

> "While the slide into general covariance turns the relativity of non-uniform motion of space-time coordinate systems into a feature general relativity shares with older theories, it does not so trivialize the relativity of the gravitational field. Even in generally covariant reformulations of these older theories, there will be an inertial field and a gravitational field existing side by side. The *unification of these two fields into one inertia-gravitational field* that splits differently into inertial and gravitational components in different coordinate systems is one of Einstein's central achievements with general relativity" ([41], 162; my italics).

Thus Galilean de-ontologization process consisted in that within the science of modernity Aristotelian 'essences' were gradually replaced by mathematical abstract objects. This is especially apparent in Newton's "*Principia*". As Sir Isaac had put it,

> "Since the ancients (as we are told by Pappus) made great account of the Science of Mechanics in the investigation of natural things; and the moderns, *laying aside substantial* forms and occult qualities, have endeavored to *subject the phenomena of nature to the laws of mathematics*; I have in this treatise cultivated Mathematics, so far as it regards Philosophy" (Newton [42], 1; my italics).

In Newton's methodology the tough demand 'to subject the phenomena of nature to the laws of mathematics' belongs to the most robust ones. One should, while contemplating the phenomena of nature, coerce his sense data in such a dry and prepared for experimental purposes way as to posit them for analytical treatment. In that way the basic mathematical abstract objects of classical mechanics, *i.e.* 'an inertial system of reference', 'a material point', 'an absolute space' and 'an absolute time', were contrived.

Correspondingly, the 'essences' of electromagnetic phenomena are represented by the abstract objects of Maxwell's equations: div **E**, rot **E**, div **H**, rot **H, j**, ρ. In Einstein's special theory of relativity the 'essences' of space and time are represented by Minkowski 4-vectors; in general theory of relativity – by second rank symmetric metric tensor g_{ij}, linked to the Riemann tensor R_{ij} and the stress-tensor T_{ij}. In quantum theory the 'essences' of the micro-processes are constituted by the wave function Ψ, representing a Gilbert space vector, while in Witten's theory the essences of space and time are 11-dimension superstrings.

It is a common place that Einstein's scientific contributions were highly motivated by the ideal of unity of physical laws, and this had a considerable influence on the whole theoretical physics community [43]. Nevertheless, all Einstein's scientific career after 1915, *i.e.* after the general theory of relativity had been achieved, was the quest for unitary theories, unification of gravitation and electrodynamics, *etc.* [44]. And it is a common place that the idea of Unity of Nature is best illustrated by *these* attempts of Einstein towards unitary theories during almost forty years than by his early works.

Yet, in my innermost conviction, Einstein's mature unification efforts and especially his general theory of relativity (GTR) are grounded on his early works and especially on his 1905 vigorous efforts to create special theory of relativity (STR) and on his audacious light quanta hypothesis (LQH). **A**s Einstein recalled later, his efforts to set up one of the basic GTR tenets – the principle of equivalence – were drawn upon his experience of creating the STR:

> "At this point, there occurred to me the happiest thought in my life [der glücklichste Gedanke meines Lebens]. ***Just as in the case with the electric field*** produced by electromagnetic induction, the gravitational field has similarly only a relative existence. For if one considers an observer in free fall, *e.g.* from the roof of a house, there exists for him during this fall no gravitational field – at least not in his immediate vicinity. Indeed, if the observer drops some bodies, then these remain relative to him in a state of rest or in uniform motion, independent of their particular chemical or physical nature" (quoted from [45], 178 ; my bold italics).

Likewise, Einstein's path-breaking 1905a paper on LQH starts with unfolding "**an *essential formal difference*** between the theoretical pictures physicists have drawn of gases and other ponderable bodies and Maxwell's theory of electromagnetic processes in so-called empty space" (my italics); the paper as a whole aims at unification of the basic research traditions of classical physics. Moreover, his 1905d paper on the STR commences with scrutinizing a "***deep asymmetry***" in the electromagnetic induction description.

Hence *the overall aim of the subsequent sections is to take the next step and to unfold the abiding influence of regulative idea of Unity of Nature on* ***all*** *Einstein's 1905 papers and especially on STR and LQH genesis and advancement.* Thus the next section of this chapter hinges upon the circle of problems that brought Einstein to electrodynamics of moving bodies. The ultimate aim of the fourth section is to answer the question: what was the train of thought that provoked

Einstein to invent defiant hypothesis of light quanta and special theory of relativity. It is contended that special theory of relativity turns out to be a meagre stage of implementation of maxwellian electrodynamics, statistical mechanics and thermodynamics reconciliation programme. The leading part in the programme was played by Einstein's 1905a light quanta paper, since the ether conception put substantial obstacles in realization of Einstein's statistical-thermodynamics design. Finally, my ultimate aim will be to underscore that the staple concept necessary to comprehend Einstein's relativity creation and all his 1905 papers *as a whole*, as well as the *order* of their arrangement was regulative idea of the systematic Unity of Nature. Moreover, it is conjectured that to conceive Einstein's STR creation and all his 1905 papers *as a whole*, as well as the *order* of their arrangement it is necessary to take into account the vicinity of Einstein's modus operandi to Kantian epistemology.

3. EINSTEIN, HERTZ, LORENTZ, POINCARÉ AND MACH

In Germany Maxwell's strenuous efforts to arrive at judicious compromise between the scientific research programmes of Young-Fresnél, Faraday and Ampére-Weber were set forth by Hermann Helmholtz and his pupil Heinrich Hertz. In Helmholtz's 1870 clear-cut paradigm charges and currents were deemed as the sources of electrical and magnetic fields. It led inevitably to H.A. Lorentz's refined dualistic worldview of the field equations and the equations of motion. Lorentz's theory was an especially ingenious blend of Maxwell's field theory and Wilhelm Weber's particle theory of electrodynamics.

Gradually Lorentz became the champion of the classical Electromagnetic Programme started by Michael Faraday and in many important aspects articulated by James Maxwell; the hard core of the programme was moulded by Newton's laws of motion and by Maxwell's equations. The substantial metaphysical component of Newtonian mechanics was the absolute space hypothesis that maintained that any inertial frame should be taken as immobile in absolute space. And, according to Lorentz, with the wave theory of light, which presupposed a medium of transmission, arose the possibility of turning the absolute space hypothesis into an empirically-testable theory. Lorentz's subtle theoretical ontology consisted of an immobile ether in which an electric charge was continuously distributed. The electrons were taken as spherical regions in the ether were electric charge and (electromagnetic) mass differ from zero. It was crucial for Lorentz's Electromagnetic Programme that an electron possesses no material mass. An electron creates an electromagnetic field which, by acting back on its source, decelerates the motion of the electron. This fundamental capacity for resisting change of motion constitutes the electromagnetic mass, which depends on the speed of the electron and accounts for the *total* inertia of the rigid bodies must be contracted by the factor $\sqrt[2]{1-(\frac{v}{c})^2}$ along the direction of

At the 1904 Saint Louis conference Poincaré accused Lorentz of having accounted for Michelson & Morley results through a whole cluster of hypotheses. This criticism was only partly justified (Zahar [52]). As a matter of fact, Lorentz had made only one additional assumption, namely the molecular forces hypothesis (MFH), which states that all forces transform like the Lorentz force (see the subsequent sections). Poincaré would have been right to contend, not that Lorentz had 'accumulated hypotheses', but that he had failed to connect the MFH with his concept of local time. Moreover, in my view it was the failure of the Electromagnetic Programme to arrive at explaining the total mass of the electron in electromagnetic way that maid the MFH especially dubious (see the next section).

Anyway Poincaré's synchronization convention helped to eliminate the puzzle. Poincaré nevertheless expected much more than that: already in "Science and Hypothesis" he took the absence of first and second order optical effects, not as accidental phenomena, but as special case of universal principle. He artfully had generalized the relativity principle by asserting that uniform motion gives rise to no detectable effects whatsoever. He boldly elevated relativity to the rank of a postulate which applies not only to mechanics but also to electromagnetism.

At the 1904 Saint Louis congress, Poincaré stressed the necessity of creating the unified theory, and it was in fact Poincaré himself who later, in 1905, constructed it [49]. Hence it was Henri Poincaré who obtained the first component of the relativity programme, namely the requirement of Lorentz-covariance. And soon it was Albert Einstein who picked up the problem after Maxwell, Helmholtz, Hertz, Lorentz and Poincaré. In 10 August 1899 'Paradies' hotel letter to Mileva Marić an ETH (Eidgenossiche Technische Hochschule) student famously confessed to his fiancée that

> "I am more and more convinced that the electrodynamics of moving bodies, as presented today, is not correct, and that it should be possible to present it in a simpler way. The introduction of the term 'ether' into the theories of electricity led to the notion of a medium of whose motion one can speak without being able, I believe, to associate a physical meaning with this statement. I think that the electric forces can be directly defined only for empty space, which is also emphasized by Hertz [...]

Electrodynamics would then be the theory of the motion of moving electricities and magnetisms in free space: which of the two conceptions must be chosen will have to be revealed by radiation experiments" (Doc. N° 52 of Einstein [53], 131).

It was Heinrich Hertz's 1890 fine paper "*Uber die Grundgleichungen der Elektrodynamik fur bewegter Körper*" that possibly turned out to be the source of the phrase "*bewegter die ElektrodynamikKörper*" in Einstein's 1905d special theory of relativity paper. Einstein used these words in the letter and thereafter to designate the complex of problems that led him to the special theory of relativity. However, Einstein was not a slavish partisan of Hertz's "*Darstellung*". From the very outset of his scientific career he had stubbornly expressed sober doubts on the role of 'des Namens Aether' in electrodynamics. Yet his skepticism was directed at Hertz's concept of the ether as a medium with a certain state of motion, and *not at the ether concept itself*. It was because Einstein attributed basic significance to the concept of 'elektrische Massen' and regarded electric currents as motions of such charges in empty space. At the start of Einstein's scientific career his views were drawn upon the lectures on electricity of his ETH physics teacher prof. H.F. Weber, as is evidently indicated by Einstein's lecture notes.

The 'substantive' concept of electricity was advanced by Wilhelm Weber and was widely accepted by the majority of German-speaking physicists, including H.F. Weber. Therein, *initially Einstein's views on electrical masses moving in the immobile ether were quite similar to the dualistic theory of Hendrik Antoon Lorentz.* Einstein concluded the abovementioned letter punctuating that the experiments were needed to choose between the two viewpoints he outlined, and his next, 10 September 1899 'Paradise' letter to Mileva mentioned an idea for experimentally investigating the influence of motion relative to the ether on light propagation in transparent bodies.

Nevertheless, Einstein's highly esteemed Her Professor of Physics showed no enthusiasm for his work, and Albert made no further mention in his correspondence of his activity in the electrodynamics of moving bodies <u>for almost two years</u>. However 'die prinzipielle Trennung von Lichtaether und Materie, Definition absoluter Ruhe' were among the important topics he vividly discussed with his close friend Michele Besso (Einstein's 4 April 1901 letter to Marić). In March 1901 Einstein wrote Mileva that he looked forward to the conclusion of "unsere *Arbeit uber die Relativbewegung*". In September 1901 he informed his boon companion Marcel Grossman on inventing a simpler method for the investigation of the motion of matter relative to ether, based 'auf gewonlichen Interferenzversuchen'. By December 1901 he was 'arbeite eifrigst' on "*die Elektrodynamik bewegter Körper*", that promised to become "eine kapitale Abhandlung" (Einstein's 17 December 1901 letter to Mileva Marić). A calculation error had earlier led him to doubt the correctness of his 'Ideen über die Relativbewegung', but he now believed in these ideas even more than ever.

He unfolded the stuff to Prof. Kleiner and the latter "thought that the experimental method proposed by me is the simplest and most appropriate and conceivable. I was very pleased with the success. I shall certainly write the paper in the coming weeks" (Einstein's letter to Marić, 19 December 1901, p. 189). Notwithstanding Prof. Kleiner's encouragement and Einstein's youthful ardor, no publication on this subject ensued for over three years – till June 1905.

Why? What was the matter? Einstein really was working hard on a 'capital memoir' on the electrodynamics of moving bodies at the end of 1901. Then he had suddenly desisted and oddly retraced to the memoir only in 1905. What did happen in the long span, and ***why had young Einstein, being initially a true adherent of the ether, became its strong enemy?***

To give a sober answer one has first to recall Einstein's derogative evaluation of his early writings - "my worthless beginner papers" [53]. All the evidence at hand indicates that the planned "*kapitale Abhandlung*" was a "far cry" from the 1905d special theory of relativity paper. On the other hand, now one knows for sure [54] that Einstein arrived at the body of results posited in his 1905d special relativity paper, in a 'sudden burst of creativity' and only *after* he had completed his first three works in the spring of 1905. *The key insight – the discovery of the relativity of simultaneity – occurred to Einstein only in late May 1905 afterthe completion of the 1905c Brownian motion paper.* In particular, when asked by the biographer Carl Seelig, Einstein enunciated:

> "Between the conception of the idea of the special theory of relativity and the completion of the corresponding published paper there passed five or six weeks" ([55], 114).

Maybe young Einstein had renounced the ether concept on finding some uncontestable, especially persuasive argument in the writings of those luminaries of science whose influence he readily and publicly admitted?

To begin with, how important was an undoubted influence of Poincaré and Mach? – For instance, in a letter to Michele Besso on 6 March 1952 Einstein famously recalled:

> "These readings were of considerable influence on my development – *along with Poincaré and Mach*" ([56], Doc. 182; my italics).

At first, how influential could occur Poincaré's eminent 'Relativity Principle', that blatantly asserted relativity of time and space? Already in 1902 Henri Poincaré enunciated that

> "*There is no absolute time.* To say two durations are equal is an assertion which has by itself no meaning and which can acquire one only by convention. Not only have we no direct intuition of the equality of two durations, but *we have not even direct intuition of the simultaneity of two events occurring in different places*: this I have explained in an article entitled '*La mesure du temps*' " (Poincaré [57], 114; my italics).

Furthermore, one of the Academia Olympia' permanent members – Einstein's boon companion Maurice Solovine – characterized Henri Poincaré's book "*La science et l'hypothese*" (first published in 1902) as one "that profoundly impressed us and kept us breathless for many weeks" (Solovine 1956; quoted from [58], 6).

Poincaré was one of the most outstanding luminaries of the fin-de-ciecle science. He left his mark on all the branches of mathematics and mathematical physics including complex analysis, topology, theory of groups, probability calculus, classical mechanics, electricity, optics, thermodynamics and astronomy.

What is more important, he can be considered as one of the true founders of special theory of relativity. Already in 1900, he had provided the operational definition clock synchronization which is commonly, though quite incorrectly, attributed to Albert Einstein (see, for instance, [59] for details). Moreover, he carefully enunciated the eminent "Lorentz covariance principle" and grounded the relativity programme on the structure of the Lorentz group [52]. Correspondingly, he had corrected the transformation rule for the electric density proposed by Hendrik Lorentz in 1904 [48]. In his later writings he also reflected on the parallel between gravity and inertia and had created one of the first Lorentz-covariant theories of gravitation [50, 51].

Yet there was a substantial point that radically distinguished Poincaré's standpoint from that of Einstein: the problem of the ether. *The consolidation of the relativity principle by Henri Poincaré did not prevent the latter from believing in ether as in the medium necessary for propagation of electromagnetic disturbances.* In my view, this can be partly explained by the fact that Poincaré's scientific advancements laid aside of the new revolutionary quantum domain. And I'll try to demonstrate in the succeeding sections, it was namely the ether conception that turned out a substantial snag for Einstein's statistical thermodynamics design, the leading part of which was played by Einstein's 1905 audacious light quanta hypothesis.

And as for Ernst Mach, the most blatant influence on young Einstein was to be exerted by the epoch-making Principle of Economy of Science according to which "Physics is Experience Arranged in Economical Order". In his eminent book "*Die Mechanik"* Ernst Mach famously argues for the overall *simplicity* and *economy* of Der Wissenschaft, and unification turns out to be a key to promote that economy and simplicity (Mach [38]).

> "One and the same view underlies both my epistemological-physical writings and my present attempt to deal with the physiology of the senses – the view, namely, that all *metaphysical elements are to be eliminated* as superfluous and as destructive of the economy of science" (Mach [60], p. XXXVIII; my italics).

An important application of the Principle of Economy of Science occurs when two mature theories, formerly separate, encounter and come into contact. For Mach this was a pivotal concern: he was driven to unify psychology and physics. At issue here was the economical requirement of needing a single orienting perspective:

> "But anyone who has in mind the gathering up of the sciences into a single whole, has to look for a conception to which he can hold in every department of science " (Mach [60], 312).

Or maybe it was pre-eminent philosopher David Hume with his profound skepticism and scathing criticism of theoretical speculations? On the 14th December 1915, Einstein admitted to a committed positivist Moritz Schlick:

> "Mach, and *even more*, Hume, whose *Treatise of Human Nature* I studied with passion and admiration *shortly before discovering the [special] theory of relativity*. Very possibly, I wouldn't have come to the solution without those philosophical studies" (quoted from [61], 247; my italics).

And much later, in a 1948 letter to Michele Besso, Einstein again stressed that

> "How far [Mach's writings] influenced my own work is, to be honest, not clear to me. In so far as I can be aware, the *immediate* influence of D. Hume on me was great. I read him with Konrad Habicht and Solovine in Bern" (quoted from Speziali [56], 153; my italics).

Yet, in my view, Mach and Hume's well-known and apparent influence should not be overestimated. Mach's search for 'economy' relations among the 'elements of experience' reflects the method of induction too strongly. And Einstein's aversion to induction is proverbial. For instance, in his 1914 inaugural address in Berlin Einstein maintained:

> "The methodology of the theoretician mandates implicitly that he use as his basis general assumptions, so-called principles, from which he can then deduce conclusions. His activity, therefore, has two parts: first, he has to *ferret out* these principles, and second, he has to develop the conclusions that can be deduced from these principles. His school provides him with excellent tools which to fulfill the second-named task [...]

But the former task, namely to establish these principles which can serve as the basis of his deductions, is one of a completely different kind. *Here there is no learnable, systematically applicable method* which would lead him to the objective. The researcher must rather *eavesdrop* on nature to become privy to these general principles by recognizing in larger sets of experimental facts certain general traits that can then be strictly and precisely formulated" (quoted from [43], 24; my italics).

Likewise, in 1919 Albert Einstein had published in the "*Berliner Tageblatt"* the essay entitled "Induction and Deduction in Physics" in which he openly maintained that "the truly great advances in our understanding of nature originated in a manner almost diametrically opposed to induction [...] The researcher does not find his system of ideas in a methodical, inductive way; rather, he smuggles up to the facts by intuitive selection among the conceivable theories that *are based upon axioms"* (quoted from [43], 42; my italics).

Furthermore, as late Einstein himself had famously recapitulated in 1949 autobiography,

> "The type of ***critical reasoning*** required for the discovery of this central point [*i.e.* the denial of absolute time, or simultaneity] was decisively *furthered*, in my case, especially by the reading of David Hume's and Ernst Mach's *philosophical* writings" (Einstein [62],53; my bold italics). Thus, David Hume and Ernst Mach's writings turned out to be especially helpful in eliminating the absolute time. But what about the ether itself?

It therefore comes at no surprise that no *direct link* between Mach's principle of economy of thought and two basic STR postulates can be traced (see section 4 for details). For instance, nowhere Einstein had contended that his postulate of the constancy of the velocity of light constitutes a *direct* consequence of the notorious Michelson-Morley experiment, not to forget his numerous declarations that he did not know about the experiment while contriving the STR. The "Light Postulate" is introduced almost parenthetically, without any discussion of its experimental grounds. For instance, in the 1905e [63] paper describing his 1905d special relativity results Einstein drops a significant phrase: "the principle of the constancy of the velocity of light used there is *of course* contained in Maxwell's equations" (Einstein 1905e [63], 172; my italics). Is it?

Rather, one can comprehend the principles of economy of thought and simplicity not in an inductivist, but in *falsificationist* way, asserting that the Lorentz-Fitzgerald contraction (LFC) hypothesis, that aimed at explaining the Michelson-Morley results within the classical physics research tradition, was an "ad hoc hypothesis".

Yet it is Elie Zahar's [52] bona fide account of ad hocness in the context of the Lorentz-Einstein transition that had convincingly exhibited that the Lorentz-Fitzgerald contraction hypothesis was not an ad hoc$_i$ (i=1,2,3) hypothesis (see the next section for details).

One gradually comes to a conclusion that the principle of economy of thought functioned just as a common motto recommending a researcher to be careful with theoretical speculations and to be as close to experience in theoretical discourse, as possible. For instance, Einstein eagerly acknowledged that Mach's critical discussion of the foundations of mechanics, which Einstein first read around 1897, helped to protect him from dogmatic "ontological" adherence for the mechanical world-view [64]. However, in 13 May 1917 letter to Besso Einstein had to admit that

> "I do not inveigh again Mach's little horse: but you know what I think about it. It cannot give birth to anything living; it can only stamp harmful vermin" ([56], Doc. 339).

In a 1974 lecture Werner Heisenberg famously recalled that Einstein confessed to him that "Perhaps I did use such philosophy [Machism] earlier, and also wrote it, but it is nonsense all the same[...]it is in fact the theory which first determines what can be observed" (quoted from [43],169).

The cases of Hertz and Hume will be more thoroughly considered further. All in all, notwithstanding Einstein's fascination by certain modicums of truth in their writings, but taking into account Einstein's aversion to inductivism, their influence also cannot be taken as decisive.

4. GENESIS, ADVANCEMENT AND FAILURE OF THE LORENTZ ELECTROMAGNETIC PROGRAMME

All the pre-maxwellian accounts of physical interactions (the pre-eminent theories of Newton, Ampére, Weber, Riemann *et al.*) were theories of interactions between several material points. Owing to Faraday and Maxwell, the Electromagnetic Field was inserted into the XIXth century physics as an element of physical reality having equal rights with the Material Point. Blatant incompatibility between the laws of mechanics and the laws of electrodynamics [65] engendered the first programme in which attempts were made to reduce electrodynamics to mechanics. Newtonian mechanics was promulgated a "fundamental theory", while electrodynamics was humbly acknowledged as a "particular" one. The ultimate aim of a reductionist programme can be achieved in two stages.

(i) Meticulous construction of basic theoretical objects of the particular theory from those of the fundamental theory. (The basic theoretical objects of Newtonian mechanics are: 'a force', 'a material point' and 'an inertial system of reference' and the basic objects of maxwellian electrodynamics embrace 'an electromagnetic field density' and ' a current density').

(ii) Scrupulous deduction of the laws of the particular theory from those of the fundamental ones.

The 'fundamental problem' of 'Maxwellian' programme consisted in construction of an ether-mechanical model capable to transmit the electromagnetic interactions *via* its elastic properties. However, this reductionist programme had failed because of two insurmountable difficulties. The first consisted in inseparable existence of both transversal and longitudinal waves in ether. The second one consisted in the queer conclusion that the ether was a solid body, as, for instance, was exhibited by Lorentz in his doctoral dissertation.

All these enabled Maxwell to begin to take the mechanical ether models as 'illustrative, but not explanatory' ones. In a pre-eminent paper especially written for '*Britannica*' and published in 1879, just before his demise, Maxwell unfolds all the proposed (beginning from Descartes's) ether models. He comes to a lamentable conclusion that all of them are unsatisfactory and that the single ether that remains is the one concocted by Huygens for light transmission.

Initially Hendrik Lorentz had commenced his scientific career by incorporating Maxwell's displacement current and the concept of electric waves into the entrenched Continental tradition of action-at-a-distance theories (see [66], pp.29-43, for details). His splendid investigations forced him to assume provisionally a dual character of physical reality. He had prudently introduced the notion of an 'electric charge' and famously completed Maxwell's equations by Newton's equations with the expression for 'Lorentz's force' that described the interaction between charge and field. Just as James Maxwell in his late years, Hendrik Lorentz was against mechanical interpretation of ether.

> "Though we all are interested in that 'state', there is no real necessity to imagine it in some way; and indeed we cannot aver much about it. However, we can imagine the existence of internal tensions in the medium surrounding a charged body or a magnet...All these was done by many physicists, and Maxwell by himself was a beginner. But there is no real necessity in it; we can advance the theory broadly and conceive a number of important phenomena without such speculations. And indeed, due to the difficulties to whom they lead, in the last years a trend occurred to avoid them all and to contrive the theory on a small number of general assumptions" (Lorentz [67]; translated from Lorentz's Russian edition, 1953, p.20).

These views were quite common among many of his contemporaries.

> "We must take the properties of ether so as we find them, investigate them with a help of the experience we have, and try to unify into a coherent picture. We should not be confused, as it is unfairly done, if it turns out that these properties differ radically from those of solid, liquid or gaseous matter"([68], p.11).

As it was stressed in Gustav Mie's popular textbook,

> "all the processes in ether can be investigated only by their actions on material bodies. These hallmarks of the physics of ether forced many physicists to abandon the very notion of ether, since it became too habitual to link with this word the ideas of somewhat material" ([69], p.10).

One of the eminent Russian physicists of the XXth century beginning emphasized that

> "the apparent simplicity of the physical properties of the void cannot be expressed better than by Greek notion 'ether'; and if because of mechanical abuses of the last times this notion is depreciated, it should not be rejected completely, but we should try to return its former sense"([70],p.10).

Note that from the very beginning Lorentz tried to unify Fresnél's ideas on the interaction of ether and matter with atomistic representations of electricity provoked by Weber and Clausius ([71], p.983). In his doctoral dissertation, completed in 1875 and devoted to light refraction, Lorentz had exhibited that longitudinal oscillations of ether (if the latter is taken as elastic liquid) should not damp. It was a ponderable reason why Lorentz preferred ether with indefinite properties. Thus, amalgamating the basic objects of both theories,

> "... Lorentz achieved the synthesis of Newton's mechanics and Maxwell's theory. The weakness of this theory consists in that it tries to conceive the phenomena by combining partial differential equations with ordinary differential equations (that describe the motion of a point). This method is unnatural. The insufficient part of the theory manifests itself in avoiding from the fact that the field existing on surface of the charge should be infinitely great. The theory was unable to explain the tremendous forces that hold the electric charges on separate particles... However, Lorentz had an idea that went beyond the boundaries of his theory. A charged body is always surrounded by a magnetic field which makes important contributions to body's inertia. Is it possible to explain the whole inertia in the electromagnetic way? It is clear that the problem can be

> solved successfully if the particles can be interpreted as regular solutions of the electromagnetic field equations..." ([72], p. 316).

Thus, Lorentz had a blatant opportunity to set forth the programme opposite to Maxwell's. Both he and Henri Poincaré saw such a programme as one offering the possibility to provide a coherent basis for all of fin-de-ciecle physics. Classical electrodynamics became the fundamental theory of this programme. Newton's mechanics was acknowledged as a particular theory. The fundamental problem of Lorentz's programme consisted in construction of electromagnetic -field model of electron in which one had to "consider the particles of matter as some local perturbations of ether"[67].

> "Indeed, one of the most important our statements will consist in the proposition that ether not only occupies all the space between molecules, atoms and electrons, but it penetrates through all these particles. We add the hypothesis that, although those particles are in motion, ether is always at rest. We can submit to this at first sight striking idea, if we consider the particles of matter as some local changes in the state of ether. This perturbations, of course, can easily move, while the elements of the medium, in which they are observed, remain at rest"([67], p.32).

This assumption unavoidably led to the proposition that in Lorentz's theory density of charge is "continuous function of coordinates, so a charged particle lacks the sharp boundary; on the contrast, it is surrounded by a thin layer in which the density of charge continuously descends from the quantity it has inside the electron, up to zero" ([67],p.32).

In Germany the same programme was implemented by E. Wiechert starting from 1894.He had maintained that the most important property of a body - its mass - should be comprehended by the action of charged particle on itself (see [55] for details).

In Great Britain sir Joseph Larmor was an active proponent of the programme described: he considered matter as consisting of rotational tensions in ether. Lines of force, electric particles, molecules, *etc.* were thought by Oliver Lodge, J.J. Thomson and other British physicists of the Victorian era to be reducible to vortices and strains in the ether.

In Russia in a harange made on 5 November 1890 in the "solemn meeting of Kazan University" Dr. Dmitry Goldhammer recounted that

> "Aspiration after revealing unity in diversity does not allow us to be satisfied with what we already have; it invites us to advance further and further, it prompts us further generalizations. Indeed, if there is an agent to whom more than a half of all the physical phenomena is obliged for their existence, will the second half be reduced to this same cause in future? And this generalization is not bolder and firmer than others. The transition from ether origin of light and electromagnetism to ether origin of usual matter with its inertia and gravitation is not bolder than the shift from two electrical and magnetic fluids to electromagnetic origin of the light ray. Not too much boldness is needed to declare now that future physics will be mechanics of ether" ([73], p.7).

But it was Wilhelm Wien who became the most passionate adherent of the reductionist programme. It was him who restlessly advocated the main purpose of physics - to bring together isolated domains of mechanics and electrodynamics. While Thomson, Maxwell, Boltzmann and Hertz strived for deriving the laws of electrodynamics from that of mechanics, Wien, beginning from 1900, tried to demonstrate that Newton's laws of motion were a peculiar case of the laws of electrodynamics. He audaciously vindicated that the first law of Newtonian mechanics is satisfied - as a consequence of the law of electromagnetic energy conservation. And the second law retained as long as it was taken to mean that the work done by a force was precisely balanced by a change in electromagnetic energy. But the third law was applicable only to forces acting between charges at rest in ether, failing for moving charges (see [55] for details).

To arrive at a more precise picture of the reductionist programme, it is necessary to quote one of the contemporaries of Lorentz and Wien:

> "Whether the inertia of matter has or has not a complete electromagnetic explanation is a question that will perhaps take many years to answer with any degree of generality. The experiments of Kaufmann seem to prove that in the case of a single electron the mass is entirely of this origin; and it is impossible therefore to avoid conclusion that at least a fraction of ordinary material inertia is also electromagnetic. Doubtless there is a psychological cause for our reluctance to accept the electromagnetic explanation as complete,

> constant familiarity with ponderable bodies having blinded us to the possibility of anything being more fundamental; but certain it is, that if we free ourselves from prejudice as much as possible and adopt the well-tried policy of choosing the simplest theory which adequately represents the phenomena, - the theory that is, which involves the least number of variables, - we must decide in favor of the complete electromagnetic explanation, which involves only the aether and its properties" ([74], p.1).

Beginning from its very genesis, the Lorentz programme did constantly advance. It had predicted a plenty of phenomena and most of its predictions were confirmed experimentally. The indubitable achievements embrace: calculation of the Fresnel coefficients (1892-1895), experiments of Roentgen (1892), Eichenwald (1903), Blondlot (1901), Wilson (1905) that went against Hertz and for Lorentz; dispersion formula (1892), Zeeman effect, law of Wiedemann-Frantz and other consequences of the electron theory of metals (1904-1905). Some of these facts were absorbed into Einstein's programme, while others had to wait for about 20 years until they were finally absorbed by quantum theory [75].

However, besides the considerations of unity, beauty, *etc*., the real necessity of carrying out the reductionist programme consisted in reconciling the Lorentz 's theory with the results of the Michelson & Morley experiment. The latter was a dismal stumbling block for Lorentz for a long time. He clearly comprehended that the only way to reconcile it with stationary ether was to suppose that interferometer arms contracted (the LFC hypothesis). Likewise,

> "in this connection, reference should be made to the proof which Hasenörl (Annalen der Physik, Band 13,p.367), reasoning from a cycling process in a moving radiating system, has given, that the second law of thermodynamics is contradicted unless either a second order contraction takes place in the direction of drift" ([76],p.72).

Nevertheless, the contraction is possible only if "the molecular forces, determining the sizes of all the bodies, propagate through ether analogously to electric forces" (the so-called MFH - Molecular Forces Hypothesis). It was quite natural for Lorentz to admit that there was no special "molecular ether" to transmit interactions between the molecules of a solid body. The intermolecular interactions propagate in good old 'electrodynamical' ether. In his LFC deduction Lorentz made use of the MFH and of his famous transformations merely. So, the MFH and consequently the LFC were as a matter of fact transitions to the

Electromagnetic World View. Hence, the validity of the MFH and of Lorentz's explanation of the Michelson & Morley experiment should be evaluated in the context of the reductionist programme. The fate of the explanation began to depend on the fate of the reductionist programme. *Lorentz's explanation of the Michelson-Morley experiment could be considered not artificial, non ad hoc, etc. only if the Electromagnetic View of Nature is correct, i.e.* only if electrodynamics is considered and treated as the pivotal, 'fundamental' theory of physics.

And the same heuristic, that made possible to explain the Michelson & Morley results, forced Lorentz and his partisans to advance. Up to 1904 the MFH was applied only macroscopically, but it was necessary to disseminate it on a single electron too. And experience seemed to support such a possibility. The experiments on 'radium rays' deflection in electric and magnetic fields were carried out already (see [55], for details). Bucherer's experiments turned out to be important also: they confirmed Lorentz's theory in the long run.

But they were theoretical investigations that appeared to play a decisive role. In 1906 Poincaré's paper [50] was published. It was convincingly demonstrated that a contractible electron can be taken as stable only under the assumption that essentially non-electromagnetic forces are invoked. However, impertinent extrapolation of the electromagnetic field transformation properties on essentially non-electromagnetic forces is a rough violation of the 'spirit' of Lorentz's programme. It is an ad hoc_3 hypothesis (in Lakatos-Zahar sense)..

Alas, Lorentz had only single option to avoid it: 'the structure less electron'. Hence the history of attempts to construct point electromagnetic electron turned out to be of extreme importance.

The opportunity of treating mass of an electron in purely electromagnetic way was discovered by Thomson in 1880. He contemplated an interaction of the charge on the surface of the spherical particle with its own electromagnetic field. Thomson revealed that this interaction engenders the force that tries to diminish the acceleration of the particle. The electromagnetic field of a charged particle moving with velocity v has a kinetic energy $T = fe2v^2/Rc^2$, where f is a form factor, R - the radius of the particle and e its charge. Consequently, one can consider $M_{em} = fe2v^2/Rc^2$ as an electromagnetic mass. The whole kinetic energy of the particle is $T = (M_0 + M_{em})\, v^2/2$.

One has a blatant opportunity to put Mo equal to zero. Thomson used it. And already in 1894 German physicist Emil Wiehert proclaimed the principles of Electromagnetic View of Nature. The ultimate reality is the ether and all the electric particles are its perturbations only. However, soon the difficulties of the 'brave new world' were revealed.

In 1903 Abraham had assiduously calculated the momentum of the electromagnetic field of moving charge $P = \int \mathbf{S}\, d^3 r$, where **S** is the Poynting vector. After messy calculations it turned out that $\mathbf{P}_{em} = (4/3)\, M_{em}\mathbf{v}$, while for usual momentum $\mathbf{p} = m\, \mathbf{v}$.

Moreover, linear and angular electromagnetic momenta entering Abraham's calculations were not proportional to their corresponding velocities, as they were in mechanics. Indeed, since they were defined by integrals all over the space, they did depend on the whole history of electron motion. It should be specially punctuated that Abraham's calculations were valid for rather peculiar case when the charge was uniformly distributed on the surface of the sphere.

But if we suppose that the charge is distributed in *all* the volume of the sphere, the result will occur no less disappointing. As it was demonstrated by Max Abraham in 1904 and Henri Poincaré in 1905 and 1906 [49, 50], deformable electron could be considered as a stable system only if ***non-electromagnetic counter-pressures*** ('Poincaré's elastics', as Richard Feynman called them in his 'Lectures') were invoked.

The luminaries of science involved in the rapidly advancing field of 'electrodynamics of moving bodies' was aware of it. When Abraham first posited that Lorentz's deformable electron required inner potential energy to preserve equilibrium, Lorentz assessed his calculations as having *extraordinary value* for his theory [77]. He had woefully admitted that if inner energy of electron would not be completely electromagnetic, the whole explanation of the Michelson-Morley experiment would become blatantly ad hoc. Abraham himself considered the necessity of introducing of nonelectric forces as a critical internal flow of Lorentz's theory. The others who were inclined to an electromagnetic viewpoint shared this estimate of Lorentz's failure. For instance, A. Bucherer agreed with Abraham that the contradictions between Lorentz's theory and pure electrodynamics were more damaging to Lorentz's theory than even the deflecting data of Kaufmann!

Now the single opportunity of further implementation of the reductionist programme consisted in the contrivance of a structure less electron model. But Lorentz's peremptory 1909 calculations said good-bye to all these hopes. The author of '*Theory of Electrons*' had calculated the force by which an electron acts on itself. If we try in the expression for the force to eliminate the structure-dependent terms by letting the radius of electron to zero, the energy of self-interaction will blatantly diverge [78]. This is meaningless from a physical point of view. The failure to construct elementary particles from field appeared to be crucial for Lorentz's programme. And in 1909 the stagnation of the reductionist programme had commenced. Subsequent artful attempts to eliminate the divergences included Paul Dirac's half-retarded and half-advanced potentials, the artful electrodynamics of Max Born and Leopold Infeld, Casimir's subtle zero-point model, *etc*. None of them succeeded. Number of unsuccessful models increased, making the ether explanation of the Michelson & Morley experiment more and more dubious and ad hoc.

Contrary to Lorentz's theory, special theory of relativity had almost nothing to maintain about the structure of electron. In 1907, in the discussion with Paul Ehrenfest, Albert Einstein explained that his Relativity Principle yielded the laws of electron motion without assuming anything about the distribution of charge within an electron. Yet it does not make the STR dubious simply because the problem of constructing electromagnetic electron is not crucial for Einstein's programme.

Nevertheless, the stubborn attempts to eliminate Lorentz's electron divergences reveal the most attractive feature of ether:

> "... So long as there is ether, there is a substructure to electromagnetism; there is a definite direction in which theory must be deepened... It seems that Lorentz lamented the absence not only of the ether concept, but of any basis by which to deepen the theory, that is to conceive of the field at an instant of time, and then to consider: what kind of things is this field; what is it made of? To respond - it is what it is, as described by the solutions to the Maxwell-Lorent--Einstein equations of electrodynamics - simply terminates the discussion. Electromagnetism is made a closed book" ([66], p.48).

5. THE INEFFICIENCY OF THE INDUCTIVIST, THE FALSIFICATIONIST AND THE LAKATOS-ZAHAR EXPLANATIONS FOR THE REASONS OF EINSTEIN'S VICTORY OVER LORENTZ

It is proverbial that Einstein's Special Theory of Relativity (STR) and Lorentz's theory co-existed at the beginning of the XX-th century as empirically-equivalent ones. In particular, both of them, though using different theoretical languages, explained the results of notorious Michelson-Morley experiment (1871,1878), as well as the results of similar experiments of Fizeau (1853), Respiga and Hook (1858), Klinkerfus (1870), Ketteler (1872), Mascare (1873), Rowland (1876), Röntgene (1885), Descoudre (1889), Lodge (1893), Rayleigh (1902), Smith (1902), Trowton (1902), Eihenwald (1903), Nordmeyer (1903), Trowton and Noble (1903), Wilson (1904), Brace (1905), Kenigsberger (1905), Strasser

(1907), *et al.* (see Boris Bursian's bona fide review [79] for details). In his 1911 conspicuous STR review Norman Campbell had specially underscored that

> "it is important to notice that there is another theory, that of Lorentz, which explains completely all the electrical laws of relatively moving systems, that the deductions from the Principle of Relativity are identical with those from the Lorentzian theory, and that both sets of deductions agree completely with all experiments that have been performed" ([80], 503).

Up to now, the various accounts of the Lorentz-Einstein transition are discussed. Their number considerably exceeds my humble abilities to meditate on all of them, hence in what follows I'll try to scrutinize the most substantial ones only. I chose inductivist and falsificationist accounts, and the explanation given by Imre Lakatos and his pupil Elie Zahar.

5.1. The Inductivist Explanation

It is famously grounded on the tough proposition that the second STR postulate (that states the invariance of the speed of light) is a *direct* generalization of the stupendous Michelson-Morley experiment.

According to Einstein's close friend and colleague Max Born,

> "the second statement, that of the constancy of the velocity of light, must be regarded as being *experimentally established with certainty*" ([81], 225; my italics).

According to an eminent philosopher of science and positivist Hans Reichenbach,

> "... it would be mistaken to argue that Einstein's theory gives an explanation of Michelson's experiment, since it does not do so. Michelson's experiment is simply taken over as an *axiom*" ([82], 20; my italics).

However, the Inductivist Explanation is apparently incorrect simultaneously from two standpoints.

Firstly, it is incorrect from the history-of-science point of view [83].

Secondly,

> "from the logical point of view, it has by now become a platitude that observation reports neither establish nor even probability high-level theories. None of Michelson's observational statements is equivalent to the proposition that in all inertial frames the speed of light is a universal constant independent of the velocity of the source" ([84], 214).

Note that too many contemporaries of Michelson and Morley were fully aware of it. For instance, in Russia a Kazan University physics Professor Dmitry A. Goldhammer maintained that

> "...in Einstein's theory the speed of light is invariant *according to the convention*" ([85], 157; my italics).

In the well-known textbook the prominent Russian physicist of the University of St. Petersburg professor Orest D. Chwolson had upholded that the second postulate

> "is an *a priori* stated axiom, a postulate on which new Weltanschauung is based. In no way one can state that it had been justified experimentally" ([86], 402; my italics).

It is rather ironic that, contrary to the inductivists like Max Born and Hans Reichenbach, the inductivists of the XXth century beginning contended that it was Lorentz's theory, and not Einstein's, that was well-established experimentally. Just to quote an eminent review of electrodynamics of the moving bodies:

> "We have described a broad range of experimental data referring to the questions of electromagnetic and optical phenomena in moving bodies, and have linked it with the development of Lorentz's theory. The latter states the existence of the ether and considers the motion of the bodies relative to it. We did observe the way according to which, following the progress of Art of the Experiment, by purely inductive

means, the theory under consideration was led to an important result, analogous to a well-known principle of classical mechanics, that in all the electromagnetic cases, as well as in purely mechanical ones, all the phenomena depend on only relative motions"([79], 35).

In 1907 Lord Rayleigh had awarded Professor Albert Michelson the Copley Medal for his "experimental investigations in optics" and had assessed the results of his experiment as follows:

> "In conjunction with E. W. Morley, he devised and carried out a very remarkable method by which, on the assumption of ether at rest, an effect depending on quantities of the order of $(v/V)^2$ would appear to be appreciable. No displacement of the rings was found. On this result the simplest explanation would be that the ether near the earth partakes fully in its orbital motion; but modern electric and optical science appears to demand a quiescent ether, and the existence of this and similar null results is fundamental for its theory"([87], 247).

And the last and, in my view, the strongest argument contra the inductivist explanation of the Lorentz-Einstein transition consists in the following. Let us recall the so-called "emission theories of light" that contested the light-constancy postulate and exchanged it with the Galilean law (that added the velocities of light and of its source).These theories [88] had no problems in explaining the Michelson-Morley result. They were specially conjured up to explain it. And they did. But they should not, if the inductivists were right.

5.2. The Falsificationist Explanation

The latter attributes Lorentz's failure and Einstein's success to their respectively ad hoc and non ad hoc responses to the results of notorious Michelson-Morley experiment.

> "An example of an unsatisfactory auxiliary hypothesis would be the Contraction Hypothesis of Fitzgerald and Lorentz which had *no falsifiable consequences* but merely served to restore the agreement between theory and experiment mainly the findings of Michelson and Morley" (Karl Popper [89], section 20; my italics).

I strongly support Imre Lakatos and Elie Zahar's stance according to which the 'ad-hocness' should be taken, in the context of Methodology of Scientific Research Programmes (or the SRP methodology), not as a property of an isolated hypothesis but as a *relation* between two consecutive theories.

> "A theory is said to be ad hoc$_1$ if it has no novel consequences as compared with its predecessor. It is ad hoc$_2$ if none of its novel predictions have been actually 'verified'; for one reason or another the experiment in question may not have been carried out, or - much worse - an experiment devised to test a novel prediction may have yielded a negative result. Finally the theory is said to be ad hoc$_3$ if it is obtained from its predecessor through a modification of the auxiliary hypothesis which does not accord with the spirit of the heuristic of the programme"([84], p.217).

To investigate the ad-hocness of the Lorentz's programme, one has to scrutinize its structure - the hard core, the heuristic and the protective belt first. According to Elie Zahar, the hard core of the Lorentz programme consisted of Maxwell's equations for the electromagnetic field, of Newton's laws of motion and of the Galilean transformations, to which Lorentz had famously added the expression for the so-called "Lorentz force". The heuristic of his programme was engendered by the following substantial metaphysical tenet: all physical phenomena are governed by actions transmitted by the ether. A theory from SRP is a conjunction of the hard core and an auxiliary hypothesis proposed according to the rules of heuristic. Lorentz's ether programme consisted of the sequence of theories L, L', L". Theory L was contrived by the conjunction of the above mentioned hard core with the following two assumptions:

a. moving clocks are not retarded;
b. material rods are not shortened by their motion through the ether.

L' was derived from L by substituting the Lorentz-Fitzgerald contraction (LFC) hypothesis for assumption (b). The LFC hypothesis states that a body moving through ether with velocity v is shortened by the factor $\sqrt[2]{1-(\frac{v}{c})^2}$

L" is the conjunction of the hard core, of the LFC and of the assumption that, contrary to (a), clocks moving with velocity v are retarded by the factor $\sqrt[2]{1-(\frac{v}{c})^2}$

Zahar convincingly exhibited that nor L', neither L" were ad hoc_1 hypotheses. First and foremost, STR, L and L' predicted different results of the Kennedy-Thorndike experiment.

Unsurprisingly L' is not an ad hoc_3 hypothesis too. Hendrik Lorentz derived the LFC hypothesis from a more profound theory - from the Molecular Forces Hypothesis (MFH): "molecular forces transform and behave like electromagnetic ones". It was quite natural for Lorentz to assume that there is no special "molecular" ether to transmit the interactions between the bodies. All the interactions should be transmitted by the common "luminiferous" ether.

Finally Zahar takes L' as non ad hoc_2, but his arguments are grounded on his well-known dubious definition of the novel fact.

> "A fact will be considered novel with respect to a given hypothesis if it did not belong to the problem situation which governed the construction of the hypothesis" ([84], 218).

Yet I approve Musgrave's ([90], pp.13-14) shrewd comment that Zahar's definition is rather dubious since it rashly puts the procedures of empirical justification from the hands of experimentalists to the hands of historians of science. However, on my humble opinion, Zahar's redefinition of the novel fact is unnecessary for the defense of L'. L' really is not ad hoc_2 but due to the other peculiar reasons. The following quotation is of importance here:

> "This assumption of shrinkage, although bold and thus far entirely hypothetical, is not impossible and is the only suggestion yet made which is capable of reconciling the negative results of second and third order experiments with quiescent ether. Poincaré (Rapports du Congres de Physique de 1900, Paris, i, pp.22-23) has raised objection to the electromagnetic theory for moving bodies, that each time new facts are brought to light a new hypothesis has to be introduced. This criticism seems to have been fairly met by Lorentz in his latest treatment of this subject" ([91], 72).

Describing and advancing Lorentz's arguments, Brace applies the startling results of Hasenörl (Annalen der Physik, 1903, band 13, p.367). Reasoning from a cyclic process in a moving radiating system, Hasenörl had exhibited that the second law of thermodynamics is contradicted unless a second order contraction takes place.

Hence not only the Michelson-Morley experiment, but all the score of the experiments establishing the second law of thermodynamics definitely support L'. This can be taken as an outstanding empirical confirmation.

It is of no wonder that many Lorentz's contemporaries had reinforced Brace's conspicuous conclusions.

> "...indeed, the possibility of contraction of the moving body in airless space seems extremely strange. But let one have a look on the essence of matter, and he will change his opinion. Indeed, according to the available information, all the forces inside the material bodies and the elastic ones in particular are the forces of the common (electromagnetic) origin" (Dmitry Goldhammer [85], 156).

Norman Campbell of Trinity College, Cambridge also took the LFC hypothesis as "artificial at the first sight. But if one considers it carefully, he can deduce, with Larmor, that the hypothesis has a firm theoretical basis. From this standpoint the negative Michelson and Morley result demonstrates that the optical properties of matter have electrical origin. Nobody doubts it now. Hence the LFC hypothesis met unanimous support and approval, and all the physicists thought that the difficulties that arose due to the Michelson-Morley experiment were solved sufficiently well" ([92], 432-434).

Paul Ehrenfest, one of the closest Einstein's friends, had also compared STR and L'. He deduced that "the critical experiment deciding in favor of one of those theories, is impossible in principle" ([93], 160).

Even Henri Poincaré himself, who had pioneered in accusing the LFC in ad hocness, admitted at 1904 St.-Louis Congress that in the light of L" the LFC was not an ad hoc hypothesis at all. In spite of the fact that LFC was initially invented to explain the results of single experiment, later it become able to explain a series of experiments.

Lorentz himself, of course, did not take the Michelson-Morley results as refutation of L'. For comparison it is instructive to consider his reaction on the results of Kaufmann's experiments. The results seemed to contradict his own theory (as well as the STR). So in a letter to H. Poincaré (8 March 1906) Lorentz confessed:

> "Unfortunately my hypothesis of the flattening of electrons is in contradiction with Kaufmann's new results, and I must abandon it. I am, therefore, at the end of my Latin. It seems to me impossible to establish a theory that demands the complete absence of an influence of translation on the phenomena of electricity and optics. I would be very happy if you would succeed in clarifying the difficulties that arise again"(quoted from [94],83).

But soon Kaufmann's results were exhibited to be erroneous.

L" or the Theory of Corresponding States was invented by Lorentz in 1904. Its genesis and advancement were described by one of his contemporaries as follows:

> "However, if we introduce this contraction into Lorentz's theory, another difficulty appears. If a piece of glass is flattened in the direction of motion, it should become double refracting for light coming perpendicular to its motion. This double refraction can be measured by light interference. But the experiments, though, did not reveal any double refraction. Again a difficulty occurred.
>
> Yet looking on his formula, Lorentz in 1904 noticed that the necessity of double refraction is eliminated, if a new assumption is conjured up. According to it, motion changes velocities of the physical processes, and, consequently, all the time intervals measured by local clocks. Any place in the moving frame of reference gets its own time. It delays from our usual time, and the delay is proportional to the velocity of the system" ([95], 173).

Thus, L" was not ad hoc_3, nor it was ad hoc_2 and ad hoc_1. Some of its novel consequences were confirmed by the experiments of Rayleigh, Brace and Bucherer on the deflections of "radium rays" in the electric and magnetic fields in 1902-1908.

5.3. The Lakatos-Zahar Splendid Explanation: Pro et Contra

Elie Zahar [84] had strikingly excavated that Lorentz research programme can be advanced to such an extent when all its empirical consequences will coincide with those of Einstein's STR. Thus, in order to make a definite choice between the two rival research programmes - Einstein's and Lorentz's – certain non-empirical criteria should be necessarily invoked. Zahar posits criteria that take into account the ***heuristic power*** of competing research programmes. Lorentz's programme had progressed up to 1905, but was eventually superseded by Einstein's programme in 1915 due to GTR (General Theory of Relativity) contrivance. Einstein's theory of gravitation famously provided successful explanations of three "critical" effects of GTR - Mercury perihelion precession, bending of light in the gravitational field and the red shift of spectral lines - that could not *in principle* be explained by the Lorentz programme.

In general, Zahar's fascinating account deals with the following three basic questions.

1. What were Einstein's sober reasons for objecting to the Classical Physics research programme and hence for starting his own?
2. Once Einstein's research programme was launched, why did other eminent scientists like Max Planck, Gilbert Lewis, Richard Tolman *et al.* work on Einstein's programme rather than on Lorentz's?
3. At what stage did the Relativity Programme empirically supersede Lorentz's?

However, in my judgement, none of Zahar's answers on the above mentioned questions is acceptable, since his rational reconstruction has the following drawbacks (unfolded by the joint efforts of historians and philosophers of science).

A. The majority of the leading physicists rejected the 'luminiferous ether' (Brace's term) already in 1910-1912 [96]. They did not wait for the creation of General Theory of Relativity.
B. Elie Zahar asserted that it was the contrivance of the General Theory of Relativity that led to elimination of the Lorentz programme. But:
   a. By or very shortly after 1915 special theory of relativity had been professionally accepted and put into use to such an extent that the general theory of relativity had not achieved even thirty or forty years (Thomas S. Kuhn);
   b. The Lorentz programme yields the constancy of c (speed of light) as a contingent fact and in this respect closer to the General Theory of Relativity than Einstein's Special Theory of Relativity where constancy of c is simply taken as a basic law [97].
C. There are some other significant discrepancies with the current history of science accounts.
   a. The scientific community of physicists shared no 'Relativity Research Programme' from 1905 to 1910: As a rule, Einstein's work had not been disentangled from Lorentz's. At the time there was only one theory in common use for the majority of scientists namely 'the Lorentz-Einstein

theory' [98]. Contemporaries of Lorentz and Einstein disputed on 'Einstein-Lorentz electron model', reflected on 'relativity principle of Lorentz and Einstein', *etc.* (Kaufmann 1906; Planck 1906; Born 1911; Jüttner 1911; Page 1912, *etc.* – see [99] and references cited therein.

b. Zahar never scrutinizes the hard core of full Einstein's programme though he often mentions a programme that contains the relativity principle together with the principle of the constancy of the velocity of light. In 1905 this programme started to degenerate while the Lorentz programme was advancing. Derivation of $E = mc^2$ was not a privilege of the Relativistic Research Programme. This equation was first obtained by Oliver Heaviside [100] and then in 1904 by Lorentz himself in "The Theory of Electrons".

c. Lorentz highly praised Einstein's reinterpretation of the equations of the Electron Theory, but never completely agreed with it. Until the end of his life he believed in ether, as well as in the notions of absolute time and absolute space [101]. However at the same time he admired the General Theory of Relativity and made an important impact to it already in 1914-1917. He did take the GTR physical space as playing the role of luminiferous ether of his electron theory. And Einstein readily agreed [102].

d. Since Zahar had defined the 'novel fact' as not belonging to the problem situation in which the new theory is contrived and proposed, he is compelled to convince the mistrustful reader that Mercury perihelion precession did *not* belong to the problem situation of the GTR creation ([103], 209). Yet history of science data cry for the opposite. Abnormal Mercury perihelion precession *did belong* to the problem situations of *all* the non-stationary theories of gravity [104, 105]. And Albert Einstein by himself had specially mentioned the perihelion precession of Mercury as a peculiar problem to be solved even in his path-breaking 1907 review, eight years *before* the GTR creation. Moreover, the correspondence between A. Einstein and M. Besso clearly indicates that Einstein asked his close friend in 1913 to compare the results of his '*Entwurf*' gravitational theory with Mercury perihelion calculations. In our times a manuscript in the hand of Michele Besso was discovered. In a multitude of pages, the first of which carries the dateline 28 August 1913, Michele Besso listed a number of problems which he had encountered while working jointly with Albert Einstein on the problem of Mercury's perihelion shift in the context of the '*Entwurf*' theory ([106], pp. 228-238). Note that the failure of the perihelion calculation was not mentioned in Einstein's subsequent 1913-1915 publications. Yet in hindsight, Einstein gave three main reasons for his rejection of the Entwurf theory the first of which consisted in *that it could not explain the perihelion shift of Mercury.*

Moreover, in 1912 G. Pavani thoroughly calculated the perihelion shift of Mercury according to Abraham's rival theory, obtaining value of 14'', 52, that is approximately one third of the observed one. Thus, Abraham's vector theory made a more accurate prediction than A. Einstein and M. Grossmann's '*Entwurf*' – the preliminary GTR version. If one takes into account that Einstein's discovery (due to Michele Besso's diligent calculations) of the failure of the '*Entwurf*' theory to yield the correct perihelion shift of Mercury was made as early as the summer 1913, one can conclude that Abraham's advancement was a vigorous spur that prompted Einstein to construct GTR out of the '*Entwurf*' (see the subsequent chapter for details).

Contemplation of **(A) -(C)** substantial shortcomings enables one to deduce that Zahar's reconstruction is rather inefficient even according to the SRP methodology liberal standards, in spite of the fact that Lakatos's SRP methodology was a firm foundation of Zahar's pre-eminent study. Indeed, the Lakatos methodology furnishes standards that apply to research programmes, not to individual theories. These standards judge the evolution of a scientific research programme over a period of time, and not its shape at a particular time. And they judge this evolution, as Paul Feyerabend had shrewdly noticed, only *in comparison* with the evolution of rivals, not by itself.

According to Imre Lakatos, all the basic methodologies of science - inductivism, falsificationism and the SRP methodology - function as historiographical (or metahistorical) research programmes and can and should be criticized by comparison of rational reconstructions that sprung out of them. But this criticism should not be taken in naive-falsificationist sense. All the methodologies should be falsified in pairs. And those one should be applauded that embraces more history-of-science data. Lakatos's excellent methodology supersedes the others because it constitutes progressive problemshift in the sequence of rational reconstructions. In Lakatos's methodology we can reconstruct as rational more aspects of scientific activity than in inductivist and falsificationist 'theories of rationality'. In their rational reconstructions scientists are either slow or irrationally rushes. However, in Elie Zahar's fine reconstruction behavior of scientists also deviates from rationality standards (though in less extent than in inductivism and falsificationism). According to E. Zahar, the scientific community was too hurry to join Einstein before 1915 and too slow to disentangle Einstein's programme from Lorentz's before 1910.

D. "Lorentz wanted to show that electromagnetic phenomena in moving bodies can be explained on the basis of a variety of processes, all of them taking place in accordance with Maxwell's equations in vacuum, referred to an all-pervading aether at rest. ***Only part*** of this research programme was superseded by R (*i.e.* Einstein's programme) in 1915. *The remainder was eventually superseded by the quantum theory* which cannot be reconciled with R. We must admit that the most comprehensive research programme in physics that existed around the turn of the century, *viz.* the programme of Lorentz, was never superseded in its entirety by another single research programme, but broke up into a variety of mutually irreconcilable programmes unless we assume that the lifework of Einstein was guided by a programme as comprehensive as that of Lorentz and unless we succeed in reconstructing its hard core. *Neither Zahar, nor anyone else has done this*" (Feyerabend [97], 26; my italics).

I reckon that at least some of the drawbacks of Zahar's thought-provoking reconstruction are engendered by the following vexed points of Lakatos's methodology.

I. The SRP methodology (or 'theory of rationality') lacks the description of the subtle mechanism of the hard cores genesis. Presumably the SRP methodology can deal only with ready-maid cores. But the lack of such an account engenders a following methodological weakness: who can guarantee that a certain rational reconstructionist had unfolded the hard core properly? The most vivid example is "*Method and Appraisal in Physical Science*" volume edited by Colin Howson. As was mordantly punctuated by Thomas Samuel Kuhn, the contributors of the volume "seldom present or even seek evidence of what actually attracted scientists to or repelled them from the various research programmes under study" ([107], 188).

Nowhere Lakatos had formulated the criteria of choice between different reconstructions of one and the same period of history of science. These criteria can be introduced by analogy with rival methodologies, though. The definite choice between different interpretations of the same hard core should be made by comparing internal historical reconstructions. However, discussing such comparisons, Imre Lakatos presupposed the existence of what he called '*actual history*', presumably the pool of all historical facts.

> "But 'actual history' of the sort Lakatos requires is a myth. The pool of all data potentially relevant to history cannot be either identified or scanned. More to the point, the data in most parts of the pool are not, until after much interpretation, the facts which appear in historical narratives. When Lakatos provides an historical case to illustrate the comparative merits of the methodology of research programmes, he

> is not selecting the elements of his internalist narrative from 'actual history' but creating them from often distant data or else choosing from the similar creations of earlier historians"(Kuhn [107],184).

It is no less important that in historiography Lakatos actually approves positivistic standpoint on the connections between theories and facts that was severely criticized by him in methodology of inductive sciences. This standpoint in history was scathingly criticized, for instance, by Robin Collingwood and Benedetto Croce, and it is superfluous to repeat it here.

II. Imre Lakatos's SRP methodology attributes an exclusive role in the evaluation of competing programmes to 'fruitfulness', but the latter is not the only criteria of choice. In my view, its importance was overestimated by Lakatos, and it distorted his view on theory-change process significantly. Furthermore,

> "though actual appraisals are for Lakatos always comparative, the single criterion in terms of which such appraisals are made is applicable to individual research programmes taken alone. Choosing between two research programmes, that is, one first locates each individually on the fruitfulness scale, the progressive/degenerative continuum, and only then does one compare the two" (Kuhn [107],189). So, the SRP methodology underestimates the subtle *dialectic* of competing research programmes. However, "historically, when a choice must be made between two research programmes, the more recent one is usually an attempt to resolve concrete difficulties encountered in the practice of its older rival. In the sciences, unlike the arts and probably also unlike mathematics, new approaches do not ordinarily emerge simply when old ones stagnate, cease to produce novelty. Instead, they await a time when the traditional approach is seen to have failed in resolving problems acknowledged to lie within its competence" (Kuhn [107], 190).

When scientists choose between the rival programmes, not only their fruitfulness is taken into account. The history of the "old" programme is also important. Could the programme really solve the well-known anomalies before the occurrence of its rival? For instance, in the case of the Lorentz-Einstein transition, many Einstein's adherents were attracted by the easiness with which the STR eliminated (yet not solved!) the quagmire of notorious difficulties of the ether theory.

III. The standards of SRP methodology are rather ineffective. They forge the description of the problem situation, in which a researcher finds himself, but they are unable to provide the advices how he can get out of the quandary. If the research programme regresses, scientist can reject it and quickly turn to the progressing one. But he has a right also, thanks to Imre Lakatos, to work within the old one. Hence the SRP methodology standards are obviously

> "arbitrary, subjective, and 'irrational'. They do not provide objective reasons for eliminating Marxism, or Aristotelianism, or Hermeticism, or for attacking new developments in the sciences. They merely indicate what critical rationalists would 'like to have' at this stage of the development of their ideology" (Feyerabend [108], 321).

The perspicuous lucid epistemological model that takes into account the critical remarks **I-III** is elucidated in Nugayev's 1999 book. So the aim of the following sections is to provide such a reconstruction of the Lorentz-Einstein transition that takes into account (A) - (D) and provides historiographical -progressive problem shift in respect to Zahar's reconstruction. I'll try to exhibit that all Einstein's 1905 writings, as well as his papers on early quantum theory and statistical mechanics, were all parts of a single research programme of synthesis of statistical thermodynamics and Maxwell's classical electrodynamics.

Lorentz tried to solve the problem of unification by reducing mechanics to electrodynamics. Inefficiency of his ether programme was unfolded by repeated failures to construct the model of pure electromagnetic electron [99]. On the contrary, holistic programme initiated by Albert Einstein to reconcile mechanics and electrodynamics was split later to two sub- programmes - quantum and relativistic - that started to advance relatively independently of each other. Zahar's account describes the advancement of the relativistic sub-programme. His fine description of an aspect of Lorentz-Einstein is, to say the least, promising. But to arrive at *deeper* comprehending for the reasons for the STR genesis, as well as for the reasons of Einstein's victory over Lorentz, one should expand the domain of history of science data substantially to embrace the early quantum history. One cannot comprehend the reasons for the STR genesis and acceptance without taking into account Einstein's works on statistical thermodynamics as well as on early quantum theory. It should be stressed that it is namely the comprehension of the advancement of Einstein's programme as a process of reconciliation of statistical mechanics, classical thermodynamics and maxwellian electrodynamics that enables one to explain internally the fact of simultaneous publication of Einstein's 1905 masterpieces.

6. WHAT WAS THE TRAIN OF THOUGHT THAT BROUGHT EINSTEIN TO LIGHT QUANTA HYPOTHESIS AND SPECIAL THEORY OF RELATIVITY CONTRIVANCE?

To give a sound answer one should first plunge into the STR paper itself (Einstein 1905d [109]). The paper commences with scrutinizing a "*deep asymmetry*" in the description of electromagnetic induction. Experience convincingly tells us that the induction current caused in the conductor by the motion of the magnet depends only on their *relative* motion. However the Maxwell-Lorentz theory provides one with *two* qualitatively different descriptions of the effect that mysteriously lead to one and the same quantitative result.

But for conceiving the true reasons of special theory of relativity genesis it is quite important to take into account that *Albert Einstein was evidently not the first to fix asymmetries in theoretical representation of the induction phenomenon.* In 1885 the asymmetries were diligently indicated by Oliver Heaviside, in 1894 – by Herman Föppl, and in 1898 – by Wielhelm Wien ([3], 377). One should especially punctuate Heinrich Hertz's fine writings. For instance, Hertz explicitly applied the term 'asymmetry' in his 1884 paper [110]. Hence namely Hertz's papers constituted part of the background to Einstein's thinking on issues in electrodynamics [111]. Indeed, at the outset of his 1905d paper Einstein invoked Maxwell's equations in their Hertzian form, namely, in the symmetrical form that Hertz presented for the first time in his 1884 fascinating paper. In his 1905d [109] STR paper Einstein is explicit about this: he appeals to the 'Maxwell-Hertz' equations. However, Hertz himself took this asymmetry as purely formal, and he simply eliminated it by re-writing Maxwell's equations in a symmetrical form. Thus the pivotal question is not how Einstein became aware of the asymmetries, but *what made them so intolerable to him.*

The key to answer the aforementioned question lies in *other* works of Albert Einstein and first and foremost in his other papers of 1905. It is proverbial that Einstein published *nothing* on the topic of optics and electrodynamics of moving bodies prior to 1905. Furthermore, it was Albert Einstein himself who had just disclosed *anotherasymmetry – and of more profound nature* – in the 1905a [112] paper "*On an heuristical point of view concerning the processes of emission and transformation of light*" that was published in the same journal "*Annalen der Physik*" but three months ***before*** the relativity paper. Look at the outset of his 1905a [112] ground-breaking paper:

> "There exist an *essential formal difference* between the theoretical pictures physicists have drawn of gases and other ponderable bodies

> and Maxwell's theory of electromagnetic processes in so-called empty space" (my italics).

And in the first part of his 1905a [112] paper Einstein excavates that joint application of mechanical and electrodynamic "theoretical pictures" for scrutinizing black-body radiation leads not only to a contradiction with experiment (his paper did not even cite the results of Lummer & Pringsheim or Rubens & Curlbaum corresponding papers), but to the glaring ***paradox*** that cannot be circumvented by common means. To exhibit it, Einstein contrives the gedanken experiment with both theories. He contemplates a cavity containing free electromagnetic field, gas molecules and Hertz's resonators. In the sequel he arrives at a conclusion that the joint application of mechanics and electrodynamics leads ***unavoidably*** to Rayleigh-Jeans law for energy density of black-body radiation. However,

> "this relation which we found as the condition for dynamic equilibrium does not only lack agreement with experiment, but it also shows that in our picture there can be no question of a definite distribution of energy between aether and matter", since "the greater we choose the range of frequencies of resonators, the greater becomes the radiation energy in space and in the limit we get $\int_0^\infty \rho_\nu \, d\nu = (R/N) (8\pi/L^3) \, T \int_0^\infty \nu^2 \rho_\nu \, d\nu = \infty$."

(Here R denotes the universal gas constant, N the number of "real molecules" in one gram-equivalent, T the absolute temperature, L the velocity of light, ν the frequency, and $\rho_\nu \, d\nu$ the energy per unit volume of that part of the radiation whose frequency lies between ν and $\nu + d\nu$).

Although it is commonly held that in the 1905a [112] paper Einstein was concerned with an explanation of the photoelectric effect, the tentative study of the paper discloses that this was not the case. The measurements of the effect at that time were not sufficiently neat to point without any doubt to a violation of classical behavior [113]. Einstein was worried not so much by the evidence dealing with photoeffect, *etc.* and appealed to fluorescence, photoelectricity and photoionization data only as to *indirect* evidence in favor of his thesis. Rather, Einstein had mostly delved into the glaring chasm, a *profound contradiction* between classical mechanics and classical electrodynamics and to the ways out of the predicament.

Japanese historian of science Tetu Hirosige [114] shrewdly attributed Einstein's sensitivity to the above mentioned inconsistencies to influence of Ernst Mach, whose writings supposedly fostered the STR inventor to out do dogmatic adherence to the mechanistic worldview. Einstein could therefore freely juxtapose Newtonian mechanics, statistical mechanics, classical thermodynamics and Maxwellian electrodynamics without reducing one to the others (see also Einstein's 1949 "Autobiography").

Jürgen Renn and Robert Schulman [115] take Einstein's anti-dogmatism as a crucial hallmark of his scientific *modus operandi* that enabled a young man of humble origin to comprehend the conceptual implications in the works of such luminaries as Lorentz, Hertz, Poincaré and Planck that they themselves were sometimes unable to discern. Unsurprisingly in their Proposal for Einstein's Membership in the Prussian Academy of Science (Berlin, 12 June 1913), M. Planck, W. Nernst, H. Rubens and E. Warburg famously contended that

> "Apart from his great productivity, Einstein has a special talent for getting to the bottom of other scientists' newly emerging views and assertions, and for assessing their *relationship to each other* and to experience with surprising certainty"([17], 338; my italics).

The other source of Einstein's unificationism is Heinrich Hertz's papers. As is well-known, Hermann Helmholtz successfully convinced Berlin Academy of Science to set up a special prize for experimental confirmation of Maxwell's theory. And it was Helmholtz's pupil -Heinrich Hertz- who got the prize in 1888. From two possible explanations of his stupendous experiments Hertz had chosen the *simplest* one [116]:

> "Helmholtz distinguishes between two forms of electric force – the electromagnetic and the electrostatic – to which, until the contrary is proved by experience, two different velocities are attributed. An interpretation of the experiments from this point of view could certainly not be incorrect, but it might perhaps be *unnecessary complicated.* In a special limiting case Helmholtz's theory becomes considerably simplified, and its equations in this case become the same as those of Maxwell's theory; only one force remains, and this is propagated with the velocity of light" ([117], 123).

Probably it was namely the attempt to justify the rationality of choosing the simplest explanation that forced Heinrich Hertz after 1888 to give up his epoch-making electromagnetic experiments and to devote the last three years of his short life to an extremely ambitious project of classical mechanics rebuilding. As he put it in his "*Principles of Mechanics*"

> "it is premature to attempt to base the equations of motion of the ether upon the laws of mechanics until we have obtained a perfect agreement as to what is understood by this name" (Hertz [118], p. XXI).

Hertz's ultimate aim was to eliminate the "force" concept. But his another aim was to reconcile the foundations of classical mechanics with positivistic Zeitgeist:

> "[...] furthermore, one would expect to find in these [electromagnetic field] equations relations between the physical magnitudes which are actually observed, and not between magnitudes which serve for calculation only" (Hertz [119], 196).

Note that the methodological principles for classical mechanics rebuilding were to be found by Hertz in Kantian epistemology. Even before he met such an eminent neo-Kantian guru as Hermann von Helmholtz, Hertz had diligently attended in a course on Kantian philosophy in Dresden. As Hertz had put it,

> "We form for ourselves images or *symbols* of external objects; and the form which we give them is such that the necessary consequents of the images in thought are always the images of the necessary consequents in nature of the things pictured. In order that these requirements may be satisfied, there must be certain conformity between nature and our thought. Experience teaches us that the requirement can be satisfied, and hence that such a conformity does in fact exist. When from our accumulated previous experience we have once succeeded in deducing images of the desired nature, we can then in a short time develop by means of them, as by means of *models*, the consequences which in the external world only arise in a comparatively long time, or as the result of our own interposition [...] *The images which we have speak of are our conceptions of things*. With the *things themselves* they are in conformity in one important

> respect, namely, in satisfying the above-mentioned requirement. For our purpose it is not necessary that they should be in conformity with the things in any other respect whatever. As a matter of fact, we do not know, nor have we any means of knowing, whether our conception of things are in conformity with them in any other than this one fundamental respect.
>
> The images which we may form of things are not determined without ambiguity by the requirement that the consequents of the images must be the images of the consequents. Various images of the same objects are possible, and these images may differ in various respects. We should at once denote as inadmissible all images which implicitly contradict the laws of our thought" (Hertz [118], 1).

As a result, Hertz's scrupulous analysis of the simplicity criterion ends with the following genuinely Kantian conclusion:

> "A doubt which makes an impression on our mind cannot be removed by calling it metaphysical; every thoughtful mind as such has needs which scientific men are accustomed to denote as metaphysical [...] It is true we cannot a priori demand from nature simplicity, nor can we judge what in her opinion is simple. But with regard to images of our own creation we can lay down requirements. We are justified in deciding that if our images are well adapted to the things, the actual relations of the things must be represented by simple relations between the images" (Hertz [118], 23).

Moreover, Hertz's Kantian background manifested itself not only in the epistemological reflection. Jed Z. Buchwald punctuated that already in 1884 Hertz had proposed such a version of Maxwell's equations that was free of the ether notion completely.

> "Hertz, one might say, wished in 1884 to remove the ether, even if Maxwell's equations were to be admitted, in order to avoid working with an entity that behaved like a laboratory object but that could not itself be directly manipulated" ([120], 278).

Furthermore, quite unlikely Maxwellian field theory, in Hertz's theoretical scheme the source of radiation continued to exist as an entity' in and of itself'. In Hertz's famous diagram the material object remains unknown, whereas the inferred field is known. This diagrammatic inversion encapsulates the originality of Hertz's physics. It was due to the fact that Hertz ignored the physical character of the object that produced his radiation – "because he boxed it in with a mental quarantine against asking questions against it – he was able to make progress where his British contemporaries had not been able to do so" ([120], 272).

Being a follower of Helmholtz, Hertz learned to watch for novel interactions between laboratory objects without worrying overmuch about the hidden processes that account for the object's effect-producing power.

Thus Hertz took the nature of electromagnetic waves a kind of "thing in itself" that admits a variety of interpretations. Any laboratory researcher chooses the version that is the simplest one to work with. The most important thing is not the interpretation but the equations depicting the relations between the objects under investigation (Hertz [121], 21).

Hertz's case suggests that the other source of Einstein's unification strategy could Kantian epistemology. It is no wonder since the papers of all the influential for Einstein scientists can be taken as being produced in the wake of neo-Kantianism.

Just to quote the Introduction of Poincaré's book:

> "the aim of science is not things themselves, as the dogmatists in their simplicity imagine, but the *relations between things;* outside those relations there is no reality knowable" (Poincaré [122], p. XIX).

Or in the same vein:

> "The object of mathematical theories is *not to reveal to us the real nature of things;* that would be an unreasonable claim. Their only object is to co-ordinate the physical laws with which physical experiment makes us acquainted, the enunciation of which, without the aid of mathematics, we should be unable to effect" (Poincaré [122], 235).

Ernst Mach's empiricism had famously sprung out of genuinely Darwinian belief that human knowledge is a necessary product of biological evolution. It was simple experience to which early organisms had responded, and it was out of such experiences that the first images of the world were eventually constructed. Eventually these constructions became *a priori*, allowing new and more subtle understandings. Furthermore, in a strict Kantian vein, Mach insisted that the human eye had a mind of its own, and consequently one perceives not direct stimuli but *relations* of stimuli. Thus one does not experience 'reality' itself but rather conceives the after effects of our nervous system's adaptation to new stimuli. 'Representational' theories of perception, positing direct correspondences between appearances and reality, were untenable for Mach.

Hence it's no wonder that Mach's empiricism drew a lot upon Kantianism. Mach credited his philosophical awakening to reading, at age 15, his father's copy of Kant's "*Prolegomena*". As he put it:

> "The book made at the time a powerful and ineffaceable impression upon me, the like of which I never afterwards experienced in any of my philosophical reading" (Mach [123], 30).

Hence one should not be surprised to learn from Jeroen van Dongen's immaculate research that "on a number of occasions Einstein actually expressed himself quite appreciative of Kant's ideas, and *some aspects of Einstein's thought did rather resemble the Kantian philosophy.* Both for instance emphasized the virtue of *striving for unity in science"* ([43], 49; my italics). And I reckon that for many conspicuous milestones of Einstein's 1905 activity Kant's influence could turn out to be crucial.

To begin with, *the very possibility* of Kant's influence on Einstein is evident: Kantian philosophy was prevalent among the educated classes in Germany in the late XIX-th century, and the extent to which it was taught in German high school was 'overwhelming' ([124], 84). Neo-Kantianism was the dominant philosophical movement in German high school from the 1870's until the First World War. Hence it is well-known that Einstein first read Kant at the age of thirteen and *again* at the age of sixteen ([125], 49). Later, being an ETH student in Zurich, he had an important opportunity to continue an acquaintance with neo-Kantianism at the lectures of August Stadler, the distinguished neo-Kantian of Marburg school ([17], 45-50). Later on Einstein was immersed in Kant many times. For instance, in his 1936 pre-eminent "*Physik und Realität*" he maintained that

> "One may say 'the eternal mystery of the world is its comprehensibility'. It is one of the greatest realizations of Immanuel

> Kant that the postulation of a real external world would be senseless without this comprehensibility" (quoted from Einstein [126], 292).

Moreover, famously reflecting on basic epistemological principles of theoretical physics, Einstein admitted that

> "the theoretical attitude here advocated is distinct from that of Kant ***only*** by the fact that we do not conceive of the categories as unalterable…They appear to be a priori only insofar as thinking without the positing of the categories and of concepts in general would be as impossible as breathing in the vacuum" (Einstein [127], 674 ;my italics).

Nevertheless, as many Einstein scholars (Michael Friedman, for instance) warned, he was a kind of an *epistemological opportunist* (see also Einstein [54], 684) for he would pick those parts of realist, positivist, conventionalist, Platonic or Kantian philosophies that he thought may be useful for justification of his scientific results and simply paid no attention to any unwelcome consequences of these philosophies. But I *contend* that when it came to ***creative***, *'regulative'* principles, Einstein, at least in 1905, derived inspiration from neo-Kantian epistemology.

Let us recall what he confidentially reported in 1918 to Max Born:

> "I am reading Kant's Prolegomena here, among other things, and am beginning to comprehend the *enormous suggestive power* that emanated from the fellow and still does" (quoted from Born [128], 25-26; my italics).

What could attract Einstein in Kantianism? -For Kant it is Man's freedom from the external World that makes science possible. The sensible world conforms to certain basic laws because the human mind ***constructs*** it according to certain laws. Constructivist foundation for scientific knowledge implies that a priori knowledge of 'things in themselves' is impossible.

Yet in the *Appendix to the "Dialectic" of the first Critique* Kant tried to provide a limited rehabilitation of the ideas of traditional metaphysics by arguing that the ideas of reason have an indispensable function in the conduct of natural science if they are understood *regulatively, i.e.* if they are taken to represent **not** metaphysical beings or entities, but rather ***goals and directions*** of scientific enquiry.

Kant famously divided the human intellect into the independent faculties of (1) sensibility, (2) understanding and (3) reason. Correspondingly, the principles governing the aforementioned faculties belong either to *constitutive* principles, or to *regulative* ones. Constitutive principles were taken as rules for the construction of the phenomenal world (*e.g.* three laws of Newtonian mechanics). The distinction between constitutive and regulative principles is drawn in Kant's first "Critique" along the following lines: the constitutive principles are those that govern the function of understanding and are necessary conditions of experience, whereas regulative principles, the 'Ideas of Reason', govern the function of reason and are not instantiated in experience in the same way. Later Hans Reichenbach contemplated constitutive principles as synthetic a priori ones and argued that to retain a role for such principles they should be *relativized* and understood as a priori only in the humble context of a particular theory.

Though constitutive principles are conditions for the possibility of the experience of the world, they do not necessitate our own particular experience of the world. Hence there is no possibility of our having an experience that directly contradicts our constitutive principles. That is why the constitutive approach cannot explain how it can be rational to abandon an established conceptual framework (a "paradigm") in favor of a new one [129].

Thus Kant asserted that the Ideas of Reason can only have a regulative rather than a constitutive role. That is why they can be used *heuristically* as a guide for our investigations, but not substantially as the actual inner principle of what we really discover. Reason demands the systematization of our knowledge, *i.e.* it *strives for Unity*. As science advances, it replaces a "narrower aspect of experience by a broader" one. Yet in experimental physics – writes Kant - even the "principles according to which we perform experiments must themselves always be derived from the knowledge of nature, and hence from the theory" (quoted from [130], p.510, n.1). Hence, due to theory-laidenness of observations, science should advance by theory unification. As neo-Kantian of Marburg school Ernst Cassirer has contended, "true unity is never thought in things as such, but in intellectual constructions" ([131], 36). In genuinely Kantian wake he stressed that unification was a purely regulative demand. And Kant himself famously contended in the *Appendixto the Dialectic* that science must take certain ideas of reason as *heuristic* ("as if") devices to encourage systematic unity.

> "The concepts of reason are, as we have said, mere ideas, and of course have no object in any sort of experience, but also do not on that account designate objects that are invented and at the same time thereby assumed to be possible. They are merely thought problematically, in order to ground regulative principles of the systematic use of the understanding in the field of experience in relation to them (as *heuristic fictions*)" (Kant [132], 659; my italics).

Along these lines, Fölsing [133] shrewdly comments that Einstein probably learned to think in terms of his "heuristic viewpoint" from his early reading of Kant's writings. Einstein's heuristic method consisted in stating, or perhaps inventing, such an assertion from which familiar facts could then be deduced. It is no wonder that Einstein's ultra-revolutionary 1905a [112] paper was entitled "*Über einen die Erzeugung und Verwandlung des Lichtes betreffenden **heuristischen** Lesichtspunkt*" ("On a ***Heuristic*** Point of View Concerning the Production and Transformation of Light").

Yet while dissociating himself from Kantian 'synthetic a priori', Einstein strongly endorses the general neo- Kantian epistemological standpoint:

> "The following, however, appears to me to be correct in Kant's statement of the problem: in thinking we use, with a certain 'right', concepts to which there is no access from the materials of sensory experiences" (Einstein [134], 22).

Correspondingly I maintain that the key notion for understanding Einstein's 1905 research activity is *Kant's profound concept of systematic **Unity of Nature*** as a regulative idea. This unity, for Kant and for Einstein, is not an ontological principle since it is meaningless to ask whether Nature in fact possesses such a unity or not. On the contrary, the idea of unity has first and foremost *epistemological* importance. Systematic unity of nature provides a *benchmark of validity for scientific hypothesis* that complements the empirical idea of confirmation.

Indeed, Kant ([132], 592) discusses the "hypothetical employment of reason", emphasizing repeatedly that the confirmation of a hypothesis by its empirical consequences can never endow such a hypothesis with universality, or 'certainty': "In natural science… there is endless conjecture, and certainty is not to be counted upon" (Kant [132], 608).

Since a given hypothesis cannot obtain the proof of its truth from 'below', from repeated experimental confirmations, something else is needed. One needs the *criterion* that can distinguish contingent and unimportant empirical generalizations from genuine fundamental Laws of Nature, which are endowed with Universality and Necessity. Kantian idea of the "truth" of a proposition is equivalent to its being a law-like statement.

> "Such concepts of reason are not created by nature, rather we question nature according to be deflective as long as it is not adequate to them [...] The hypothetical use of reason is therefore directed at the *systematic unity* of the understanding's cognition, which, however, is the *touchstone of truth* for its rules" (Kant [132], 592; my italics).

Hence from the multitude of different uniformities only those can be taken as having law-like necessity that can be fitted into a unified, systematized general system.

Compare with Einstein's dictum:

> "A system has truth-content according to the certainty and completeness of its coordination-possibility to the totality of experience. A correct proposition borrows its 'truth' from the truth-content of a system to which it belongs" (Einstein [135], 13).

What I want to stress is that it was this 'holistic' stand that enabled Einstein already in 1906 to disregard the results of Kaufmann's "crucial" experiments (that seemed to corroborate the Abraham-Bucherer theory) and to refute the "Lorentz-Einstein" theory ([136], 253 [137], 124).

As Einstein had put it, the rival theories (*e.g.* Abraham's electron theory)

> "Have rather small probabilities, because their fundamental assumptions (concerning the mass of moving electrons) are not explainable in terms of theoretical systems which embrace a greater complex of phenomena" (Einstein as quoted in [136], 253).

Einstein readily admitted that any domain of phenomena can be described by a host of empirically equivalent alternative theories. However if for conventionalist like Pierre Duhem such an underdetermination of theory by facts meant that there can be no ultimately true theory, the Einsteinian approach in the Kantian wake not only provides a meaning to the regulative ideal of a final theory. It also illuminates Einstein's remarks that despite this underdetermination at any given time there is *only one* correct theory: *the theory with the greatest power of unification* (Einstein [138], 226).

Thus Einstein's attraction in the 1905a [112] paper to the subject of theory of quanta was provoked by its *unifying possibilities*, for its capacities to arrive at a *fusion* of Maxwell's electrodynamics and Boltzmann's statistical thermodynamics. So he starts the paper with the heart of what troubled him most – the Rift, the glaring chasm in the foundations of physics that was felt most sharply in Lorentz 's "Theory of Electrons".

Later, in his pre-eminent "*Physik und Realität*", Einstein had eloquently described the problem situation as follows:

> "... Lorentz achieved the synthesis of Newton's mechanics and Maxwell's theory. The weakness of this theory consists in that it tries to conceive the phenomena by combining partial differential equations with ordinary differential equations (that describe the motion of a point). This method is unnatural. The insufficient part of the theory manifests itself in avoiding from the fact that the field existing on surface of the charge should be infinitely great. The theory was unable to explain the tremendous forces that hold the electric charges on separate particles... " ([139], p. 316).

How had Einstein intended to deal with the profound contradiction in his 1905a epoch-making paper?

While considering Einstein's solution, one should take into account that *all* Einstein's papers from 1902 to 1905 have one hallmark in common: statistical-thermodynamics approach. Thomas S. Kuhn had famously punctuated that *what brought Einstein to idea of photon was a coherent advancement of a research program started in 1902*, a programme "so nearly independent of Planck that it would almost certainly have led to the black-body law even if Planck had never lived" ([140], 171). From the start of his scientific career Einstein was "deeply impressed", as Martin Klein shrewdly punctuated, by 'the simplicity and scope of classical thermodynamics'. Yet for him thermodynamics necessarily included the statistical approach he had learned from Ludwig Boltzmann's writings. So he began to unfold statistical thermodynamics and the result was a series of three papers published in 1902, 1903 and 1904 [143, 144]. It should be stressed that *namely they provide the clue for apprehending his 1905a [112] paper on quanta, his 1905b dissertation [141], 1905c [142] paper on Brownian motion and 1905d [109] paper on special relativity.*

The first significant result consisted in that for physical systems of extraordinary general sort Einstein has produced, by the summer of 1903, both a generalized measure for temperature T and entropy S, that contained some universal constant χ. By the time he finished his 1903 [143] paper, Einstein had recognized that χ could be evaluated in terms of the values of the gas constant and of Avogadro's number. But the theory that had led him to the constant was, however, applicable to systems far more general than gases. It should therefore have a correspondingly general physical foundation. The basis should reflect statistical-mechanical nature of the approach that led him to the constant, comprehending not only its role as a scale factor for temperature, but also its position as a multiplier in the probabilistic definition of entropy. Physical significance of χ was the central quandary attacked in Einstein's third statistical paper "*On the General Molecular Theory of Heat*", submitted to "*Annalen der Physik*" in the spring of 1904. The solution of stubbornly recalcitrant problem consisted in taking into account the phenomena of energy fluctuations. Einstein elucidated that $\varepsilon^{-2} = 2\chi\ T\ dE/dT$, where ε^{-2} is a measure of thermal stability of the system. And it was comprehension of the physical sense of the constant that directed his attention to the black-body problem.

> "The equation just found would permit an exact determination of the universal constant χ if it were possible to determine the energy fluctuation of the system. In the present state of our knowledge, however, that is not the case. Indeed, for only one sort of physical system can we presume from experience that an energy fluctuation occurs. That system is empty space filled with thermal radiation" (Einstein [144], 360; translated by Thomas Kuhn in [140]).

At least one more step in the programme of statistical thermodynamics advancement was needed, and Einstein took it in the ground-breaking 1905a [112] paper. Its content reveals that Einstein had begun to seek a black-body law of his own, that he had quickly elucidated the paradox, evinced in the contradiction between statistical mechanics and maxwellian electrodynamics, and that he had provisionally dropped the search for the law in favor of an investigation of the paradox itself. This is apparent from the very beginning of his already quoted paper (translated in [113]). The first part of the 1905a [112] paper ended by revelation of the so-called 'ultraviolet catastrophe'. How did Einstein resolve the paradox?

In the second part of his 1905a [112] paper Einstein applies thermodynamics, statistical mechanics and maxwellian electrodynamics to peer at the domain of empirical reality covered by Wien's radiation law. Einstein takes $\beta = h/k = N\, h/R$ (R denotes the universal gas constant, N the number of "real molecules" in one gram-equivalent, h is Planck's constant and k is Boltzmann's constant) as undefined constant in 1905a [112] paper and hence he writes $R\beta/N$ everywhere instead of h. Joint application of the three mature theories enables Einstein to arrive at apparently deductive argument: if monochromatic radiation of frequency ν and energy E is enclosed in the volume V_0, then the probability W that at any moment all the radiation energy will be discovered in the partial volume V of the volume V_0 is given by

$$W = (V/V_0)^{E/h\nu} \quad \textbf{(i)}$$

Yet in the same paper Einstein had previously exhibited that in the case of n independently moving particles enclosed in a volume V_0 the probability of finding them all momentarily in the sub volume V is

$$W = (V/V_0)^{n} \quad \textbf{(ii)}$$

Comparing equations (i) and (ii), Einstein draws a conclusion that "*monochromatic radiation of small density behaves in thermodynamic respects as though it consists of distinct independent energy quanta of magnitude hv*".

Thus, the upshot that radiation in the cavity consists of independent energy quanta follows ***directly*** from application of general principles of thermodynamics and statistical mechanics to radiation phenomena.

But in 1905 all the available experimental data, relevant to fluorescence, photoelectricity and photoionization data, provided only *indirect* evidence in favor of quantum hypothesis. Hence, to check the audacious and queer hypothesis of quanta, Einstein was forced to perform a "crucial experiment" of a very peculiar, freaky kind. He had to compare the quantum results with the results of another entrenched, 'old' theory contrived independently of the 1905a light quanta hypothesis. It is important that this theory had to be sufficiently 'old' to accumulate the results of many experiments. So, if the 1905a [112] paper results had matched the results of fairly different theory, that sprung out of different problem situation, they would have provided an especially reliable verification of the light quanta hypothesis. Let us recall: "A system has truth-content according to the certainty and completeness of its coordination-possibility to the totality of experience. A correct proposition borrows its 'truth' from the truth-content of a system to which it belongs" (Einstein [135], 13). In the opposite case the 1905a theory would have 'falsified' not by a single 'critical experiment' but by a whole multitude of the well-established experimental data.

Thus the next Einstein's result (1905b [141]) turned out to be especially crucial for the 1905a [112] verification. In the 1905b [141] paper Einstein assiduously worked out the principles of Brownian motion that were *directly* verified by Perrin's experiments. The importance of the 1905b [141] paper for the 1905a [112] one was promulgated by Einstein much later. On 17 January 1952 he confessed to his old friend and colleague Max von Laue:

> "When one goes through your collection of verifications of the special relativity theory, one believes that Maxwell's theory is firmly established. But in 1905 I knew *already with certainty* that it leads to the wrong fluctuations in radiation pressure, and consequently to an incorrect Brownian motion of a mirror in a Planckian radiation cavity" (quoted from [54], 177; my italics).

It is important to note that the abovementioned Brownian motion results, quite independently of Einstein, were obtained by Marian Smoluchowski from Krakow University ([145, 146]) and by William Sutherland [147].This fact indicates the fundamental value and importance of Einstein's results and that again Einstein managed to get to the mainstream of the world physics. Let us recall again that in their Proposal for Einstein's Membership in the Prussian Academy of Science, M. Planck *et al.* had shrewdly emphasized that

> " Einstein has a special talent for getting to the bottom of other scientists' newly emerging views and assertions, and for assessing *their relationship to each other* and to experience with surprising certainty" ([45], 338; my italics).

Yet this blatant for 1905 Einstein result was posited to the scientific community only in 1909 when Einstein applied his theory of Brownian motion to a two-sided mirror immersed in thermal radiation. He exhibited that the mirror would be unable to carry out a Brownian motion indefinitely, if the fluctuations in the radiation pressure on its surfaces were solely due to the effects of random waves, as predicted by Maxwell's theory. But only the stubborn presence of an additional term, corresponding to pressure fluctuations due to the impact of random particles, guarantees the continued Brownian motion of the mirror. Einstein shrewdly unfolded that similar fluctuation terms in the energy were consequences of Planck's law. He took such fluctuation phenomena as the ***strongest argument*** for ascribing physical significance to the mysterious light quanta. Only after the "crucial experiment", that is only *after* the 1905b [141] paper could Einstein look forward for investigating the consequences of his light quantum hypothesis, and so he returned to his half-forgotten "unsere Arbeit uber die Relativbewegung", eine "kapitale Abhandlung".

So far, so good.

> "if the monochromatic radiation (of sufficiently small density) in the sense of entropy dependence upon volume behaves itself as a discontinuous medium, consisting of energy quanta Rβv/N, a question occurs: if they are not the laws of creation and conversion of light such as if it consists of similar energy quanta?" (Einstein 1905a [112], 236).

That is the important question put up by Einstein at the end of § 6 of his 1905a [112]. But ***the ether conception turned out to be a substantive snag***. *It stubbornly hampered positive answer and put obstacles in uncoiling Einstein's statistical-thermodynamics approach.* Indeed

> "mechanical and purely electromagnetic interpretations of optical and electromagnetic phenomena have in common that in both cases electromagnetic field is considered as a special state of hypothetical medium filling all the space. Namely in that point two interpretations mentioned differ radically from Newton's emission theory, in which light consists of moving particles. According to Newton, space should be considered as possessing neither ponderable matter, nor light rays, *i.e. absolutely empty*" (Einstein 1905a [112], 236; my italics).

To *contrive* a quantum theory of radiation, one needs electromagnetic fields as *independent* entities that can be emitted by the source 'just as in Newton's emitting theory' (*i.e.* energy transmitted in a process of emission should not be dissipated in space, but should be completely preserved until an elementary act of absorption). However, within the Lorentz programme an electromagnetic field is considered as a specific state of ether - a state of medium that is *continuously* distributed in space. In such a medium an elementary radiation act is necessarily connected only with a spherical wave.

Nevertheless, aversion to ether and acceptance of emission theory should lead to Walter Ritz's 1908 'ballistic hypothesis': velocity of quantum should depend on the velocity of its source. In Ritz's theory velocity of light is not constant, but is equal to v+ c, where v is a relative velocity of the observer and the source. But Einstein, by contrast, never intended to down Maxwell's theory, just as Newton, the inventor of the emission theory, had not rejected the wave theory 300 years earlier. In the 1905a photon paper Einstein had to make the following reservation:

> "Wave theory operating with point continuous functions is excellently justified when describing purely optical phenomena and perhaps would not be replaced by another theory" (Einstein 1905a [112], 237).

In the Lorentz research programme this predicament was absent. Indeed, in the reference frame that is at rest relative to ether light propagates with constant velocity *independent of the velocity of the source.* Hence, if one tries to give up the idea of luminiferous ether, but to come to terms with Maxwell's theory at the same time, s/he should disown ballistic hypothesis and postulate a special "principle of constancy of velocity of light"(**I**). Later, in April of 1922, Einstein had confessed to Viscardini:

> "I rejected this [emission] hypothesis at that time, because it leads to tremendous theoretical difficulties (*e.g.* the expectation of shadow formation by a screen that moves relative to the light source)" (quoted from [54], 182).

The second STR basic principle - "the principle of relativity"(**II**) - follows immediately from the tenet that there is no luminiferous ether and, consequently, no absolute system of reference.

The two postulates, **(I)** + **(II)**, the relativity principle plus the principle of constancy of velocity of light, are quite sufficient, according to Einstein, to create the electrodynamics of moving bodies. Yet, since "the theory based on these two principles should not to lead to contradictory results, one must renounce the customary rule of addition of velocities" (Einstein [148], 125).

And namely that was exhibited in the 1905d [109] paper "*On the Electrodynamics of Moving Bodies*", published several months ***after*** the photon paper. Einstein had dug out the hidden assumption - the basic premise of the Galileo addition law - that the statements of time, as well as of the shapes of moving bodies have the sense independent of the state of motion of the reference frame. He unfolded that the acceptance of the "principle of relativity" together with the "principle of constancy of light" is equivalent to modification of the simultaneity concept and to clock delay in moving reference frame. Note that Albert Einstein was not an idle 'philosopher-scientist' contemplating the essences of Space and Time in the Bursts of Divine Inspiration. He was *forced* to elevated philosophical reflections on the nature of space and time by his common *research practice*, by a mundane physical problem of reconciling Principle of Relativity with the Light Constancy Postulate (see [149] for details) .

Hence, at least in that case, Einstein's application of Hume and Mach's philosophical writings was "*highly selective*" ([149], 359). His ultimate goals were not so much to apprehend Hume's and Mach's refined philosophical reflections as to find in them concrete ideas that may be useful in his mundane research practice. Thus Hume's and Mach's writings only provided "the type of critical reasoning" (Einstein [150]) necessary for the clearing the thinking from the remnants of obsolete metaphysical systems "rooted unrecognized in the unconscious" [150].

Nevertheless, the momentum, the driving force, the *mainspring* had to be found on the other ways. In a letter to Besso on 6 January 1948 Einstein admitted that

> "I see his [Mach's] weakness in this, that he more or less believed science to consist of mere 'ordering' of empirical 'material'; that is to say, he did not recognize the ***freely constructive element*** in the formation of concepts. In a way he thought that theories arouse through discoveries and not through inventions. He even went so far that he regarded 'sensations' not only as a material which has to be investigated, but, as it were, as the building blocks of the real world" ([151], Doc 153; translated by Gerald Holton; my bold italics).

As for sage David Hume, Mathias Slavov stressed that Hume's and Einstein's conceptions of space and time have *substantial* differences [152]. In Hume's adamant epistemological doctrine, space and time are *direct* abstractions from simple perceptions. On the contrary, Einstein constantly emphasized that the basic concepts of science are "***free creations***" of the human mind. In that respect Einstein's views were evidently closer to Kant. So, the positive drive for creative *Drung* could be found in Kant's ***constructivist*** foundation for scientific knowledge that restricted science to the realm of appearances stating that a priori knowledge of things in themselves is impossible. Much later Einstein had to admit:

> "I did not grow up in the Kantian tradition, but *came to understand* the truly valuable which is to be found in his doctrine, alongside of errors which today are quite obvious, quite late. It is contained in the sentence: 'The real is not given [gegeben] to us, but put to us [aufgegeben]' [by way of a riddle]" (Einstein [150], 680; quoted from [153]).

Kant took even mathematics – maintained to be most stable and certain because of its being analytical – as an *a priori synthetic judgement.* As he famously stressed in "Critique" and "*Prolegomena"*, the essential feature of pure mathematical cognition, differentiating it from all other a priori cognition, is that it must throughout proceed not from concepts, but always and only through the ***construction*** of concepts. Because pure mathematical cognition, in its propositions, must therefore go beyond the concept to that which is contained in the *intuition* corresponding to it, its propositions can and must never arise through the analysis of concepts, *i.e.* analytically, and so are one and all synthetic.

Accordingly, on a fine Christmas day of 1919 Einstein published in the '*Berliner Tageblatt'* the abovementioned essay "*Induction and Deduction in Physics"* in which he confessed that

> "The truly great advances in our understanding of nature originated in a manner almost diametrically opposed to induction. The *intuitive* grasp of the essentials of a large complex of facts leads the scientist to the postulation of a hypothetical basic law, or select such basic laws [...] [The researcher] does not find his system of ideas in a methodical, inductive way; rather, he snuggles up to the facts by ***intuitive*** selection among the conceivable theories that are based upon axioms" (quoted from [43], 41; my italics).

The Kantian tenet of the intuitive character of mathematics means the limiting of mathematics to those objects that are constitutable [Konstruierbar]. 'Intuitive' is equal to 'constitutable'. As Wittgenstein has later put it in genuinely Kantian terms, "But the mathematician is not a discoverer, he *is* an *inventor"*.

It is not accidental that Kant contemplated objectivity of science as resulting from the manner in which the manifold of sensibility was ordered under the categories of the understanding by means of spatial and temporal categories. This is why mathematics could so effectively describe objective reality for Kant: mathematical constructs are related to the pure intuitions of space and time. And this is why natural science must be mathematical.

Hence mathematical statements are true in virtue of their application in experience to exhibit the behavior of empirical bodies. While mathematical judgements are obtained through *construction* in *pure intuition*, they count as cognitions only because they are necessary connected to experience in the sense that geometrical space was contemplated as a condition of appearance.

> "Although we know a priori in synthetic judgements a great deal regarding space in general and the figures which productive imagination describes in it, and can obtain such judgements without actually requiring any experience, yet even this knowledge would be nothing but a playing with a mere figment of the brain, were it not that space has to be regarded as a condition of the appearances which constitute the material for outer experience.

Those pure synthetic judgements therefore relate, though only mediately, to *possible experience*, or rather to the possibility of experience; and upon that alone is founded the objective validity of their synthesis" (Kant [132], 196; my italics).

In a sense the abstract objects of a theory are constituted by the laws of a theory. And objectivity is connected not to the existence of *things* but to the *objective validity of relations*. Accordingly, in the 1905a [112] paper, constructing the mathematical abstract object "light quanta" out of the joint basis of maxwellian electrodynamics and statistical thermodynamics, Einstein was bothered not with grasping the 'essences' of radiation phenomena but with the problems of ***reconciling*** the interrelations of different classical research traditions, *i.e.* maxwellian electrodynamics, statistical mechanics and thermodynamics. Let us recall again that in their Proposal for Einstein's Membership in the Prussian Academy of Science, M. Planck *et al.* had shrewdly emphasized that

> "Einstein has a special talent for getting to the bottom of other scientists' newly emerging views and assertions, and for assessing *their relationship to each other* and to experience with surprising certainty" ([45], 338; my italics).

As Einstein recalled later, 'The real is not given [gegeben] to us, but put to us' (Einstein 1949 [127],680), *i.e.* it is 'constructed' due to our research activity.

In a nutshell, Einstein's Kantianism can be characterized by a motto of another prominent neo – Kantian of Marburg school - Wilhelm Windelband: "Kant verstehen, heißt über ihn hinausgehen". It is proverbial that Einstein's philosophical evolution after the General Relativity,*i.e.* after 1915 carried him further and further from Humean and Machian empiricist bias toward Neo-Kantian tradition represented by Weyl, Eddington, Cassirer, Husserl *et al.* and the mathematical speculative methodology embodied in a sequence of unified theories. Thus I am not contending here that Einstein of 1905 was a true (neo) Kantian, trying to implement the abstract and murky tenets of "*Critique*" into his mundane unification practice. Yet I insist that *the seeds of Einstein's late methodology lie in his 1905 activity* connected with his fruitful efforts to reconcile maxwellian electrodynamics and statistical thermodynamics.

Well, if all abovementioned is true, a question necessarily arises: *why Einstein in the 1905d special relativity paper did not cite his 1905a paper on light quanta?*

To give a judicious answer one has to dwell into Einstein's 1905 correspondence. Writing to his boon companion Conrad Habicht in 1905 and sending him the fruits of his labors at that time, Einstein characterized his light quanta paper as "*very revolutionary*", while he humbly noted that the relativity paper might be "interesting in its kinematical part". So, *reference in the paper, introducing significant changes mainly of metaphysical character, on the hypothesis that had already introduced ultra-revolutionary changes and had apparently contradicted respectable and well-established and entrenched Maxwell's theory, could hardly make the arguments stronger.*

At the first Solvay Congress Einstein himself had to admit "provisional character of this concept [light quanta] which does not seem reconcilable with the experimentally verified consequences of the wave theory" (quoted from [154], 884). The problem situation was even worse since *direct experimental evidence* in favor of light quanta was absent. It famously appeared only in 1923 (the Compton effect).

Being taken independently, the special theory of relativity did not explain any *new* experimental fact. Predictions of the Lorentz theory were identical to that of the STR, so that it seemed not be possible to distinguish between these theories on experimental grounds. Furthermore, most of Einstein's contemporaries had scrutinized the "Lorentz-Einstein electron model", reflected on the "principle of relativity of Lorentz and Einstein", *etc*. At the time of publication of Lorentz's second order theory (1904) the only experimental data available to test the both theories were Kaufmann's measurements of the masses of slowly moving electrons. But they were initially comprehended as contradicting *both* the STR and the Lorentz theory. It took almost a year for Einstein to answer on Kaufmann's paper. One can only imagine how the special theory of relativity was evaluated by the scientific community in 1905 - 1906!

Moreover, Einstein did not expose the connections between 1905a [112] and 1905d [109] until 1909. However, *without these connections the STR postulates can be evaluated as ad hoc hypotheses. And they were.* (The reaction of Henri Poincaré and of the French school is the most blatant example). So, being confronted with many rivals, why did Einstein preferred special theory of relativity? What undisguised advantages did it offer over the theories of Lorentz, Ritz and others?

The answer leads one to Einstein's Kantian bias once more. The Einsteinian approach in the Kantian vein illuminates Einstein's seemingly puzzling remarks that despite this undetermination at any given time there *is* only one correct theory: the theory with the greatest power of unification at that time (Einstein [138]; see [124] for details).

One is usually told that in constructing special theory of relativity Einstein had invented a "*theory of principle*", rather than a "*constructive theory*". Yet things are not that simple.

Indeed, it was Einstein himself who made a profound distinction between 'principle' theories and 'constructive' ones. Constructive theories try to "build up a picture of the more complex phenomena out of the materials of a relatively simple formal scheme from which they start out" (Einstein 1919 as quoted from [43],49). An example of a constructive theory is kinetic theory that successfully attempts at reducing mechanical and thermal properties of gases to movements of molecules.

On the contrary, principle theories do not start out from hypothetical constructions, but rather from empirically ascertained principles.

> "Thus the science of thermodynamics seeks by analytical means to deduce necessary conditions, which separate events have to satisfy, from the universally experienced fact that perpetual motion is impossible". In *explicitly Kantian* terms Einstein in 1919 distinguishes between the abovementioned kinds of theories: "principal theories employ the *analytic*, not the *synthetic* method" (quoted from [43], 50; my italics).

Prima facie it is to his close friend Michele Besso that Einstein dedicated the only acknowledgement in his 1905d paper, the paper that stands out for its *lack of any reference to the literature*. Furthermore, in the 1905d [109] paper "the failure of attempts to detect a motion of the earth relative to the 'light medium'" is used as evidential support only for *one of the two* basic postulates – for the "Principle of Relativity". The "Light Postulate" is introduced almost parenthetically, without any sober discussion of its experimental grounds. Only in the 1905e [155] paper, while commenting on the 1905d [109] results, Einstein drops a phrase: "the principle of the constancy of the velocity of light used there is *of course* contained in Maxwell's equations" (Einstein [156], 172). But for him the 1905d [109] paper was only a provisional construct, only a milestone in implementing the unification programme. Einstein contended that

> "a physical theory can only be satisfactory, if its structures are composed of elementary foundations. The theory of relativity is just as little ultimately satisfactory as, for example, classical thermodynamics was before Boltzmann had interpreted the entropy as probability"(Einstein to Arnold Sommerfeld on 14 January 1909; quoted from [151], 10).

Thus, the statement that 1905d [109] paper constituted a theory of principle is merely half of the truth. In reality the 1905d [109] theory was a constructive one that only *posited itself* as a theory of principle (possibly due to tactical reasons for he tried to save the STR from the mordant criticism directed against the light quanta). That is why two years later, trying to explain the STR foundations to broad physical community, Einstein humbly described his relativity theory as "an attempt to summarize the studies that have resulted to date from the *merger* of the H.A.Lorentz's theory and the principle of relativity" (Einstein [156], 253).

But the situation could not last over a long period of time. Einstein had to throw his cards up and to disclose the link between his 1905a and 1905d papers four years later. In 1909, in Salzburg, he made a report at the 81st meeting of German

Natural Scientists and Physicians under the promising heading "*On the Development of our Views on the Nature and Structure of Radiation"*. It represented practically the first attempt to comprehend all his works as a whole. And it was one of the first public reports of the STR inventor dedicated to diligent explanation of its foundations. The report starts with a succinct recapitulation of luminiferous ether theory that ends by startling phrase: "However, today we must regard the ether hypothesis as an obsolete standpoint". Why? – What I want to emphasize is that for the answer Einstein dwells *not to the Michelson-Morley or Fizeau experiments*, but elucidates that

> "It is even undeniable that there is an extensive group of facts concerning radiation that shows that light possesses certain fundamental properties that can be understood far more readily from the standpoint of Newton's *emission theory* of light than from the standpoint of the wave theory. It is therefore my opinion that the next stage in the development of theoretical physics will bring us a theory of light that can be understood as a kind of *fusion* of the wave and emission theories of light"(Einstein [157], 379; my italics).

And the above mentioned experiments are brought into consideration only in the context of the "cardinal aspect in which the electromagnetic theory agrees with, or, more accurately, *seems to agree* with the kinetic theory" (ibid).

7. CONCLUSIONS

The pivotal claim to put forward is that to conceive many important facets of Einstein's special theory of relativity and the light quanta hypothesis creation and all his 1905 papers ***as a whole*** as well as the ***order*** of their presentation one should resort to Einstein's strenuous efforts to reconcile maxwellian electrodynamics and statistical thermodynamics. And to comprehend the importance of the latter one should appeal to neo-Kantian epistemology.

Nevertheless, this is not to assert that Einstein was a committed Neo-Kantian of Baden or Marburg schools. No. In a letter, written three weeks after the creation of General Relativity, Einstein applauded Moritz Schlick's scathing critique of Kant:

> "Truly masterful is your position to the doctrine of Kant and his followers. The trust in the '*apodictic certainty*' of 'synthetic a priori judgements' is already heavily undermined if one realizes the

> invalidity of even just one of these judgements" (14 December 1915, A. Einstein to M. Schlick, quoted from [43], 46; my italics).

All in all, Einstein's 'true' overall philosophical standpoint was *eclecticism*. But I insist that the most important Kantian concept necessary to comprehend Einstein's 1905 research activity is Kant's regulative idea of the systematic Unity of Nature.

The basic problem of the chapter is what was the train of thought that brought Einstein to his special theory of relativity and bold hypothesis of light quanta. Hence the first question to answer is why Einstein, being initially a true adherent of the luminiferous ether, became its strong enemy. To give a sober answer one has to turn to the outset of the STR paper. It famously starts with unfolding the "deep asymmetry" in the description of electromagnetic induction. It is important that Albert Einstein was by no means the first to note asymmetries in theoretical representation of the induction phenomenon, though. Hence the pertinent question is not why Einstein became aware of asymmetries and contradictions, but *what made them so intolerable to him.*

On my view, the key answer to the questions posed lies in *other* works of Albert Einstein and first and foremost in his 1905 papers. It was Albert Einstein himself who had acutely revealed the basic asymmetry - of more profound nature – in the 1905a [112] paper "*On an heuristical point of view concerning the processes of emission and transformation of light*" that was published in the same journal "*Annalen der Physik"* but three months *before* the paper on special relativity. In §1 of this thought-provoking paper Einstein unfolds that joint application of mechanical and electrodynamic "theoretical pictures" for description of black-body radiation leads not only to contradiction with experiment, but to the queer ***paradox*** that cannot be circumvented by common methods. Einstein's attraction to the subject of theory of quanta was provoked by its *unifying possibilities*, for its capacities to arrive at a *unification* of maxwellian electrodynamics and Boltzmann's statistical thermodynamics and to eliminate the glaring chasm between the dominant paradigms of Classical Physics. Hence he starts the paper with the heart of what troubled him most – the *profound rift* in the foundations of physics that was felt most sharply in Lorentz's Electron Theory (see also [99]). How did Einstein intend to eliminate the contradiction?

To answer the question one should immerse himself in Einstein's first papers published in the "*Annalender Physik"*. *All* of them have one hallmark in common: statistical thermodynamics approach. Thomas S. Kuhn had revealed that *what brought Einstein to idea of photon was a coherent advancement of a research program started in 1902*. Thus, the conclusion that radiation in the cavity consists of independent energy quanta follows directly from application of general principles of thermodynamics and statistical mechanics to processes of radiation.

Yet all the experiments available in 1905 provided only *indirect* evidence in favor of quantum hypothesis. Hence, to try the audacious hypothesis of quanta Einstein had to appeal to the *"crucial experiment"* of the very subtle kind: to compare the quantum results with the results of another entrenched theory build up independently of the 1905a theory. It is important that this theory should be sufficiently 'old' to accumulate the results of many experiments. So, if the 1905a paper results had coincided with the results of fairly different theory, they would have provided an especially reliable verification. It was namely that point where Kant's idea of the systematic Unity of Nature was applied. "A system has truth-content according to the certainty and completeness of its coordination-possibility to the totality of experience. A correct proposition borrows its 'truth' from the truth-content of a system to which it belongs" (Einstein [135], 13). In the opposite case the 1905a theory would have been 'falsified' not by a single 'critical experiment' but by a multitude of the well-established experimental data.

Thus the next 1905b paper turned out to be crucial for verification of the 1905a [112] one. In the 1905b [141] Einstein worked out the principles of Brownian motion directly verified by Perrin's experiments. Only after this "crucial experiment", that is only *after* the 1905b [141], corroborated by Smoluchowski and Sutherland's *independent* calculations, could Einstein look forward for investigating the consequences of his light quantum hypothesis, and so he returned to his forlorn "eine kapitale Abhandlung". Indeed,

> "if the monochromatic radiation (of sufficiently small density) in the sense of entropy dependence upon volume behaves itself as a discontinuous medium, consisting of energy quanta Rβν/N, a question occurs: *if they are not the laws of creation and conversion of light such as if it consists of similar energy quanta*?" (Einstein 1905a [112], 236; my italics).

That is the question put up by Einstein at the end of one of his 1905a [112] paragraphs. *But the luminiferous ether conception fiercely hampered the positive answer and put insurmountable obstacles in realization of Einstein's statistical-thermodynamics programme.*

Hence the ether concept had to be abandoned once and for all and the special theory of relativity grounded on two postulates should be set up. The two postulates - the Relativity Principle and the Principle of Light Constancy - were quite sufficient, according to Einstein, to create the electrodynamics of moving bodies. Yet, for "the theory based on these two principles should not to lead to contradictory results, one must renounce the customary rule of addition of velocities" (Einstein [148], 125). And namely that was performed in "*On the Electrodynamics of Moving Bodies*", published several months *after* the photon paper.

To recapitulate, Einstein was undoubtedly influenced by Hume, Mach, Poincaré, Hertz *et al.*, that is reflected in many important documents that embrace private communications, public lectures, *etc.*, relating to different periods of his life. However, if one dwells into his scientific papers themselves to comprehend his *modus operandi*, one finds out sober reasons to believe that *actually*, at least in 1905, in his actual *research practice*, he had held an epistemological position very close to neo-Kantian epistemology. And the most important Kantian concept necessary to comprehend Einstein's 1905 research activity appeared to be Kant's regulative idea of the systematic Unity of Nature.

REFERENCES

[1] D. Siegel, *Innovation in Maxwell's Electromagnetic Theory: Molecular Vortices, Displacement Current, and Light.* Cambridge University Press: Cambridge, 1991.

[2] M. Morrison, *Unifying Scientific Theories: Physical Concepts and Mathematical Structures.* Cambridge University Press: Cambridge, 2000. [http://dx.doi.org/10.1017/CBO9780511527333]

[3] O. Darrigol, *Electrodynamics from Ampere to Einstein.* Oxford University Press: Oxford, 2001.

[4] S. D'Agostino, Maxwell's Dimensional Approach to the Velocity of Light: Rise and Fall of a Paradigm. *In: Mathematical Models and Physical Theories.* Academia Nazionale Della Scienze: Rome, 1984, pp. 147-167.

[5] L. Campbell, and W. Garnett, *The Life of James Clerk Maxwell.* Macmillan: London, 1882.

[6] G. Hon, and B.R. Goldstein, "Maxwell's contrived analogy: An early version of the methodology of modeling", *Stud. Hist. Philos. M.P.,* vol. 43, no. 4, pp. 236-257, 2012. [http://dx.doi.org/10.1016/j.shpsb.2012.07.001]

[7] J.C. Maxwell, "On Faraday's Lines of Force", *In: Transactions of the Cambridge Philosophical Society*, vol. X, part 1, read on December 10, 1855 and on February 11, 1856 and published in 1858. Reprinted in: *The Scientific Papers of James Clerk Maxwell*, 1890, vol. 1., Cambridge: Cambridge University Press and New York: Dover, 1952, pp. 155-229.

[8] J.C. Maxwell, "A Dynamical Theory of Electromagnetic Field", *Philos. T. Roy. Soc. A*, vol. 155, pp. 459-512, January 1865. Reprinted in: *The Scientific Papers of James Clerk Maxwell*, 1890, vol. 1, Cambridge: Cambridge University Press and New York: Dover, 1952, pp. 526-597

[9] J.C. Maxwell, On Action at a Distance", *P.R. Inst. GB*, vol. VII, pp.48-49, 1873-1875. Reprinted in: *The Scientific Papers of James Clerk Maxwell*, 1890, vol. 1, Cambridge: Cambridge University Press and New York: Dover, 1952, pp. 315-320

[10] B. Hunt, *The Maxwellians. Cornell.* Cornell University Press: Ithaca, 2005.

[11] B. Mahon, *The Man Who Changed Everything. The Life of James Clerk Maxwell.* John Wiley:

London, 2003.

[12] A. Chalmers, "Maxwell, Mechanism and the Nature of Electricity", *Phys. Perspect.,* vol. 3, no. 4, pp. 425-438, 2001.
[http://dx.doi.org/10.1007/PL00000539]

[13] H. Hertz, ""On Electromagnetic Waves in Air and Their Reflection", Wiedemann's Annalen, vol. 34, pp.609-623, February 1888. Translated in:Heinrich Rudolph Hertz. Electrical Waves", *L.: Macmillan,* vol. 1893, pp. 124-136, 1888. b

[14] H. Hertz, *Electric Waves.* Macmillan: London, 1893.

[15] J.Z. Buchwald, "Reflections on Hertz and Hertzian Dipole." *In: Heinrich Hertz: Classical Physicist, Modern Philosopher.* Kluwer: Dordrecht, 1998, pp. 269-280.
[http://dx.doi.org/10.1007/978-94-015-8855-3_15]

[16] H. Hertz, "On the Relations between Maxwell's Fundamental Electromagnetic Equations and the Fundamental Equations of the Opposing Electromagnetics", *Wiedemann's Annalen*, vol.23, pp.84-103, September 1884.Translated in: *Heinrich Hertz. Miscellaneous Papers.* L.: Macmillan and Co., 1896, pp.273-290.

[17] I. Shapiro, "On the History of the Discovery of the Maxwell equations", *Sov. Phys. Usp.,* vol. 15, no. 5, pp. 651-659, 1973.
[http://dx.doi.org/10.1070/PU1973v015n05ABEH005038]

[18] W. Whewell, *The Philosophy of Inductive Sciences, founded upon their history.* John W. Parker and Son: London, 1847.

[19] L. Boltzmann, "Ueber Faraday's Kraftlinien, von James Clerk Maxwell" *In: Herausgegeben von. L. Boltzmann.* W. Engelmann: Leipzig, 1895, pp. 315-320.

[20] W. Hamilton, *Discussions on Philosophy and Literature, Education and University Reform.* Mc Lachlan and Stewart: Edinburgh, 1853.

[21] G. Graham, "Scottish philosophy in the XIX-th Century." *In: Stanford Encyclopedia of Philosophy.* Stanford University Press: Stanford, 2011.

[22] I. Kant, "Prolegomena to any future metaphysics that will be able to come forward as science (Transl. by Gary Hatfield)" *In: Immanuel Kant. Theoretical Philosophy after 1781* Cambridge University Press: Cambridge, 2002, pp. 29-170.

[23] I. Kant, *The Critique of Pure Reason.* The Pennsylvania State University Press: Pennsylvania, 2010.

[24] J.C. Maxwell, "Hermann Ludwig Ferdinand Helmholtz", *Nature*, vol. XV, pp. 389-391, March 1877. Reprinted in: *The Scientific Papers of James Clerk Maxwell*, 1890, vol. 2, Cambridge: The University Press and N.Y.: Dover, 1952, pp. 592-59.

[25] J. Mertz, *A History of European Thought in the Nineteenth Century*, 4 vols, Edinburgh: William Blackwood and Sons:, 1903-1912. Vols 1 and 2 reprinted as "*A History of European Scientific Thought in the Nineteenth Century.* N.Y.: Dover, 1964

[26] R. Olson, *Scottish Philosophy and British Physics, 1750-1880: A Study in the Foundations of the Victorian Scientific Style.* Princeton University Press: Princeton, NJ, 1975.

[27] P. Harman, *The Natural Philosophy of J.C. Maxwell.* Cambridge University Press: Cambridge, 2001.

[28] J.C. Maxwell, "On Physical Lines of Force". *Philos. Mag.*, Series 4, vol. 21, pp. 161-175, 281-291, March and April 1861; *Philos. Mag.*, Series 4, vol. 23, pp. 12-24, 89-95, May 1861, February 1862. Reprinted in: *The Scientific Papers of James Clerk Maxwell*, 1890, vol. 1, Cambridge: The University Press and N.Y.: Dover, 1952, pp. 451-513

[29] E. Whittaker, *A History of the Theories of Ether and Electricity: From the Age of Descartes to the Close of the Nineteenth Century. L.* Longmans, Green and Co.: N.Y, 1910.

[30] R.M. Nugayev, *Reconstruction of Mature Theory Change: A Theory –Change Model.* Peter Lang:

Frankfurt am Main, 1999.

[31] D. Sengupta, and T. Sarkar, "Maxwell, hertz, the maxwellians and the early history of electromagnetic waves", *IEEE Antennas Propag. Mag.*, vol. 45, no. 2, pp. 12-19, 2003. [http://dx.doi.org/10.1109/MAP.2003.1203114]

[32] G. Fitzgerald, "On the electromagnetic theory of the reflection and refraction of light". *Proc. Roy. Soc.*, vol. 171, pp. 691-711, January 1880; reprinted in: *The Scientific Writings of the late George Francis Fitzgerald.* London: Longmans, Green & Co, 1902, pp. 45-72

[33] R.M. Nugayev, "The Ptolemy-Copernicus Transition: Intertheoretic Context", *Almagest,* vol. 4, no. 1, pp. 96-119, 2013. [http://dx.doi.org/10.1484/J.ALMA.5.100985]

[34] R. Feynman, R. Leighton, and M. Sands, *The Feynman Lectures on Physics.* 2nd ed. Addison Wesley: N.Y., 2005.

[35] A. Koyré, *From the Closed World to the Infinite Universe.* John Hopkins Press: Baltimore, 1957.

[36] E.A. Burtt, *The Metaphysical Foundations of Modern Science.* Dover Publications: N.Y., 2003.

[37] E. Mc Mullin, "Galilean idealization", *Stud. Hist. Philos. Sci.,* vol. 16, no. 3, pp. 247-273, 1985. [http://dx.doi.org/10.1016/0039-3681(85)90003-2]

[38] E. Mach, *The Science of Mechanics: A Critical and Historical Account of its Developments.* (Translated by T. J. McCormmach). La Salle: Open Court, 1960

[39] E. Husserl, *The Crisis of European Sciences and Transcendental Philosophy.,* D. Carr, Ed., Northwestern University Press: Evanston, 1970.

[40] M. Kline, *Mathematics and the Search for Knowledge.* Oxford University Press: N.Y., 1985.

[41] M. Janssen, "The twins and the bucket: How Einstein made gravity rather than motion relative in general relativity", *Stud. Hist. Philos. Sci. Part Stud. Hist. Philos. Mod. Phys.,* vol. 43, no. 3, pp. 159-175, 2012. [http://dx.doi.org/10.1016/j.shpsb.2012.01.003]

[42] I. Newton, *The Mathematical Principles of Natural Philosophy (translated into English by Andrew Motte. With preface of Mr. Roger Cotes).* Daniel Ades: New York, 1846.

[43] J. Van Dongen, *Einstein's Unification.* Cambridge University Press: Cambridge, 2010. [http://dx.doi.org/10.1017/CBO9780511781377]

[44] V.P. Vizgin, *Unified Field Theories in the first half of the 20th century.* Springer: Basel, 2011. [http://dx.doi.org/10.1007/978-3-0348-0174-4]

[45] A. Pais, *Subtle is the Lord. The Science and Life of Albert Einstein.* Oxford University Press: Oxford, 1982.

[46] H.A. Lorentz, "De relative bewegung van de aarde en den aether", *Versl. Kon. Ak. Wet.,* vol. 1, pp. 74-79, 1892.

[47] H.A. Lorentz, *Versuch einer Theorie der elektrischen und optischen Erscheinungen in bewegter Korpern.* Brill: Leiden, 1895.

[48] H.A. Lorentz, "Electromagnetic phenomena in a system moving with any velocity less than that of light", *Proc. Acad. Sci. Amst,* vol. 6, pp. 809-831, 1904.

[49] H. Poincaré, "Sur La Dynamique de l'Electron", *C.R. Acad. Sci. Paris,* vol. 140, pp. 1504-1508, 1905.

[50] H. Poincaré, "Sur La Dynamique de l'Electron", *Rend. Circ. Mat. Palermo,* vol. 21, no. 12, pp. 9-175, 1906.

[51] H. Poincaré, "La theorie de Lorentz et le Principe de Reaction", *Arch. Neerland,* vol. 5, no. 2, pp. 252-278, 1900.

[52] E. Zahar, *Einstein's Revolution: A Study in Heuristic La Salle: Open Court,* 1989.

[53] *Albert Einstein / Mileva Marić. The Love Letters*. Ed. and with an introduction by Jürgen Renn and Robert Schulmann. Transl. by Syawn Smith. Princeton NJ: Princeton University Press

[54] R. Rynasiewicz, "The Construction of the Special Theory: Some Queries and Considerations." *In: Einstein. The Formative Years, 1879-1909.* Birkhauser: Boston, 2000, pp. 159-201. [http://dx.doi.org/10.1007/978-1-4612-0311-7_6]

[55] C. Seelig, *Albert Einstein. Leben und Werk Eines Genies Unserer Zeit.* Europa Verlag: Leipzig, 1960.

[56] P. Speziali, Ed., *Albert Einstein-Michele Besso: Correspondence.* Hermann: Paris, 1972.

[57] H. Poincaré, *[1902] Science and Hypothesis. London and Newcastle-on-Tyne.* The Walter Scott Publishing Co., 1905.

[58] D. Howard, J. Stachel, Ed., *Einstein. The Formative Years, 1879-1909.* Birkhauser: Boston, 2000. [http://dx.doi.org/10.1007/978-1-4612-0311-7]

[59] A.A . Logunov, *On the articles by Henri Poncaré "On the Dynamics of the Electron.* Dubna: JINR preprint, 2001

[60] E. Mach, *The Analysis of Sensations and the Relation of the Physical to the Psychical.* (Translated by C.M. Williams). La Salle: Open Court, 1984

[61] M. Slavov, "Empiricism and Relationism Intertwined: Hume and Einstein's STR", *Theoria,* vol. 31, no. 2, pp. 247-263, 2016. [http://dx.doi.org/10.1387/theoria.14846]

[62] A. Einstein, "Remarks Concerning the Essays Brought together in this Co-operative Volume." *In: Albert Einstein.* vol. 1-2. Philosopher-Scientist: Evanston, IL, 1949, pp. 665-688.

[63] A. Einstein, "Ist die Trägheit eines Körpers von seinem Energienhalt abhängig?", *Ann. Phys.*, vol. 18, pp. 639-641, September 1905. English translation of Anna Beck as "Does the Inertia of a Body Depend on its Energy Content?", In: *The Collected papers of Albert Einstein. Vol. 2. The Swiss years: writings, 1900 – 1909.* Princeton, New Jersey: Princeton University Press, 1989, pp. 172 - 174.

[64] S. Saunders, and H. Brown, "Reflections on Ether." *In: The Philosophy of Vacuum.* Clarendon Press: Oxford, 1991, pp. 27-64.

[65] R. Nugayev, "The history of quantum theory as a decisive argument favoring einstein over lorentz", *Philos. Sci.,* vol. 52, pp. 44-63, 1985. [http://dx.doi.org/10.1086/289221]

[66] S. Saunders, and H. Brown, "Reflections on Ether." *In: The Philosophy of Vacuum.* Clarendon Press: Oxford, 1991, pp. 27-64.

[67] H.A. Lorentz, *The theory of electrons and its applications to the phenomena of light and radiant heat.* G.F. Stechart & Co: New York, 1909.

[68] P. Lenard, *Aether and Matter. A lecture read in Heidelberg Academy of Science at 4 June 1910.* Russian Academy of Science Publishers: St. Petersburg, 1910 (in Russian).

[69] G. Mie, *Lecture Course on Electricity and Magnetism.* Mathesis: Odessa, 1912. (in Russian)

[70] O.D. Chwolson, "The principle of relativity", *J. Russ. Phys. Chem. Soc.,* vol. 12, pp. 377-409, 1912. [in Russian].

[71] S. Goldberg, "The lorentz theory of electrons and einstein's theory of relativity", *Am. J. Phys.,* vol. 37, pp. 982-994, 1969. [http://dx.doi.org/10.1119/1.1975220]

[72] A. Einstein, "Physik und Realität", *J. Franklin Inst.,* vol. 221, pp. 121-164, 1936.

[73] D.A. Goldhammer, *What We Know About the Aether.* Kazan University Press: Kazan, 1890. (in

Russian)

[74] D.F. Comstock, "The relation of mass to energy", *Philos. Mag.,* vol. 15, pp. 321-336, 1908.

[75] P.K. Feyerabend, "Zahar on Einstein", *Br. J. Philos. Sci.,* vol. 25, pp. 25-28, 1974. [http://dx.doi.org/10.1093/bjps/25.1.25]

[76] D.B. Brace, "The negative results of second and third order tests of the 'Aether Drift' and possible first order methods", *Philos. Mag.,* vol. 10, pp. 72-85, 1905.

[77] H.A. Lorentz, "Weiterbildung der Maxwellschen Theorie. Elektrontheorie", In: *Encyklopedie der Mathematischen Wissenschaften.* Band 5/2,Teubner.: Leipzig, 1904.

[78] F. Rohrlich, *Classical Charged Particles.* Addyson-Wesley: New York, 1973.

[79] B. Bursian, "Experimental Investigations Studying the Influence of Motion of Matter on Ether." *In: New Ideas in Physics.* vol. Vol. 3. Russian Academy of Science Press: St.-Petersburg, 1904, pp. 30-45. (in Russian)

[80] N.R. Campbell, "The common sense of relativity", *Philos. Mag.,* vol. 21, pp. 112-126, 1911.

[81] M. Born, *Einstein's Theory of Relativity.* Dover: New York, 1962.

[82] H. Reichenbach, *The Philosophy of Space and Time.* Dover: New York, 1958.

[83] I. Lakatos, "Falsification and the Methodology of Scientific Research Programmes." *In: Criticism and the Growth of Knowledge.* Cambridge University Press: London, 1970, pp. 91-195. [http://dx.doi.org/10.1017/CBO9781139171434.009]

[84] E. Zahar, *Einstein's Revolution: A Study in Heuristic.* La Salle: Open Court, 1989.

[85] D.A. Goldhammer, "New ideas in modern physics", *Fizicheskoye Obozreniye* (Review of Physics), vol. 12(1-2), pp. 1-35, 76-155, January-February 1911 (in Russian)

[86] O.D. Chwolson, "The principle of relativity", *J. Russ. Phys. Chem. Soc.,* vol. 12, pp. 377-409, 1912. [in Russian].

[87] L. Rayleigh, "Anniversary address", *Proc. Roy. Soc. A,* 80(538), pp. 239-251, April 1908

[88] R.C. Tolman, "Some emission theories of light", *Phys. Rev.,* vol. 35, pp. 136-143, 1912.

[89] K. Popper, *Logik der forschung.* Springer: Wien, 1938.

[90] A. Musgrave, "Logical versus historical theories of confirmation", *Br. J. Philos. Sci.,* vol. 25, pp. 1-23, 1974. [http://dx.doi.org/10.1093/bjps/25.1.1]

[91] D.B. Brace, "The negative results of second and third order tests of the 'Aether Drift' and possible first order methods", *Philos. Mag.,* vol. 10, pp. 72-85, 1905.

[92] N.R. Campbell, *Modern Electrical Theory.* St. Petersburg: Izdatelstvo Obrazovaniye. (Education Publishing House), 1912 (in Russian).

[93] P. Ehrenfest, "Crisis in the Hypothesis of Light Ether", *J. Russ. Phys. Chem. Soc.,* vol. 4, pp. 151-162, 1913. [in Russian].

[94] A.I. Miller, *Albert Einstein's Special Theory of Relativity.* Addison-Wesley: London, 1981.

[95] D.A. Goldhammer, "Time. Space. Aether", *J. Russ. Phys. Chem. Soc.,* vol. 5, pp. 173-189, 1912. [in Russian].

[96] K.F. Schaffner, "Einstein versus Lorentz: research programmes and the logic of comparative theory evaluation", *Br. J. Philos. Sci.,* vol. 24, pp. 45-78, 1973.

[97] P.K. Feyerabend, "Zahar on Einstein", *Br. J. Philos. Sci.,* vol. 25, pp. 25-28, 1974. [http://dx.doi.org/10.1093/bjps/25.1.25]

[98] A.I. Miller, "On Lorentz's Methodology", *Br. J. Philos. Sci.,* vol. 24, pp. 29-45, 1973.

[99] R.M. Nugayev, *Reconstruction of Mature Theory Change: A Theory-Change Model.* Peter Lang: Frankfurt am Main, 1999.

[100] B. Bolotovsky, "The birth and development of new knowledge: Oliver Heaviside", *Int. Stud. Philos. Sci.,* vol. 1, no. 2, pp. 143-175, 1987. [http://dx.doi.org/10.1080/02698598708573289]

[101] N. Nersessian, "Why wasn't Lorentz Einstein?", *Centaurus,* vol. 29, pp. 205-242, 1986. [http://dx.doi.org/10.1111/j.1600-0498.1986.tb00857.x]

[102] R. Mc Cormmach, "H.A. Lorentz and the Electromagnetic View of Nature", *Isis,* vol. 61, pp. 459-497, 1970. [http://dx.doi.org/10.1086/350681]

[103] J. Illy, *Einstein teaches Lorentz, Lorentz teaches Einstein.* Hungarian Academy of Science Publishers: Budapest, 1981.

[104] I.F. Subbotin, "Theory of relativity and Celestial Mechanics", *Astron. J.,* vol. 23, pp. 251-263, 1956. [in Russian].

[105] V.P. Vizgin, *Relativistic theory of gravity: genesis and development.1900-1915.* Moscow: Nauka: (in Russian).

[106] J. Renn, and T. Sauer, "Pathways out of Classical Physics: Einstein's Double Strategy in his Search for the Gravitational Field Equations." *In: The Genesis of General Relativity.* vol. 1. Springer: Dordrecht, 2007, pp. 113-312.

[107] T.S. Kuhn, "The Halt and the Blind: Philosophy and History of Science", *Br. J. Philos. Sci.,* vol. 31, pp. 181-192, 1980. [http://dx.doi.org/10.1093/bjps/31.2.181]

[108] P.K. Feyerabend, "Critique of Scientific Reason." *In: Method and Appraisal in the Physical Sciences. The Critical Background to Modern Physics, 1800-1905.* Cambridge University Press: Cambridge, 1976, pp. 309-339.

[109] A. Einstein, "Zur Elektrodynamik bewegter Körper", *Ann. Physik,* vol.17, pp. 891-921, June 1905. English translation in: The Principle of Relativity, New York: Dover: 1923. See also the translation of Anna Beck in: The Collected papers of Albert Einstein. Vol. 2. The Swiss years: writings, 1900 – 1909. Princeton, New Jersey: Princeton University Press, 1989, pp. 140 – 171.

[110] H. Hertz, "Ueber die Beziehungen zwischen den Maxwell'schen elektrodinamischen Grundgleichungen und den Grundgleichungen der gegnerischen Elektrodynamik", *Ann. Phys.,* vol. 23, pp. 84-103, 1884. [http://dx.doi.org/10.1002/andp.18842590904]

[111] G. Hon, and B.R. Goldstein, "How Einstein Made Asymmetries Disappear: Symmetry and Relativity in 1905", *Arch. Hist. Exact Sci.,* vol. 59, no. 11, pp. 437-544, 2015.

[112] A. Einstein, "Über einen die Erzeugung und Verwandlung des Lichtes betreffenden heuristischen Gesichtspunkt", *Ann. Phys.*, vol. 17, pp. 132-48, March 1905. English translation by Anna Beck in: *The Collected papers of Albert Einstein. Vol. 2. The Swiss years: writings, 1900 – 1909.* Princeton, New Jersey: Princeton University Press, 1989, pp. 86-103.

[113] D. Ter Haar, *The Old Quantum Theory.* Pergamon Press: Oxford, 1967.

[114] T. Hirosige, "The ether problem, the mechanistic world view and the origins of the theory of relativity", *Hist. Stud. Phys. Sci.,* vol. 7, pp. 3-82, 1976. [http://dx.doi.org/10.2307/27757354]

[115] J. Renn, and R. Schulmann, "Introduction." *In: Albert Einstein/ Mileva Maric. The Love Letters.* Princeton University Press: Princeton, 1992, pp. 3-22.

[116] R. Smirnov-Rueda, "Were Hertz's "crucial experiments" on propagation of electromagnetic

interaction conclusive?" *In: Instantaneous Action at a Distance in Modern Physics: pro and contra.* Nova Science Publishers: N.Y., 2010, pp. 57-69.

[117] H. Hertz, "The forces of Electric Oscillations, Treated According to Maxwell's Theory." *In: Heinrich Rudolph Hertz. Electric Waves. L.* Macmillan, 1893, pp. 137-159.

[118] H. Hertz, *The Principles of Mechanics Presented in a New Form* (translated by D.E. Jones). L.: Macmillan, 1899.

[119] H. Hertz, "On the Fundamental Equations of Electromagnetics for Bodies at Rest." *In: Heinrich Rudolph Hertz. Electric Waves. L.* Macmillan, 1893, pp. 195-240.

[120] J. Buchwald, *The creation of scientific effects: Heinrich Hertz and electric waves.* The University of Chicago Press: Chicago, 1994. [http://dx.doi.org/10.7208/chicago/9780226078915.001.0001]

[121] H. Hertz, *Electric Waves. L.* Macmillan, 1893.

[122] H. Poincaré, *[1902] Science and Hypothesis. London and Newcastle-on-Tyne.* The Walter Scott Publishing Co, 1905.

[123] E. Mach, *The Analysis of Sensations and the Relation of the Physical to the Psychical.* (Translated by C.M.Williams). La Salle: Open Court, 1984

[124] M. Beller, "Kant's Impact on Einstein's Thought." *In: Einstein: The Formative Years, 1879-1909.* Birkhauser: Boston, 2000. [http://dx.doi.org/10.1007/978-1-4612-0311-7_4]

[125] D. Howard, "Einstein, Kant and the Origins of Logical Positivism." *In: Language, Logic and the Structure of Scientific Theories: The Carnap -Reichenbach Centennial.* University of Pittsburgh Press: Pittsburgh, 1994, pp. 45-105.

[126] A. Einstein, *Ideas and Opinions.* Crown Publishers, Inc.: New York, 1954.

[127] A. Einstein, "Remarks Concerning the Essays Brought together in this Co-operative Volume." *Albert Einstein.* vol. 1-2. Philosopher-Scientist: Evanston, IL, 1949, pp. 665-688.

[128] M. Born, *The Born-Einstein Letters: Correspondence between Albert Einstein and Max and Hedwig Born from 1916 to 1955. I. Born translation.* Walker and Company: N.Y., 1971.

[129] J. Everett, "The constitutive a priori and the distinction between mathematical and physical possibility", *Stud. Hist. Philos. M. P.,* vol. 52, pp. 139-152, 2015. [http://dx.doi.org/10.1016/j.shpsb.2015.05.004]

[130] G. Buchdahl, *Metaphysics and the Philosophy of Science. The Classical Origins: Descartes to Kant.* Cambridge, Massachusetts: MIT Press, 1969 [46] *Albert Einstein / Mileva Marić. The Love Letters.* Ed. and with an introduction by Jürgen Renn and Robert Schulmann. Transl. by Syawn Smith. Princeton NJ: Princeton University Press, 1992.

[131] E. Cassirer, *Substance and Function and Einstein's Theory of Relativity.* (Translated by W.C. Swabey and M.C. Swabey). Repr. Chicago: The Open Court Publishing, 1923.

[132] I. Kant, *The Critique of Pure Reason. Repr. (Translated by Paul Guyer and Allen W. Wood),* P. Guyer, A.W. Wood, Eds., Cambridge University Press: Cambridge, 1998. [http://dx.doi.org/10.1017/CBO9780511804649]

[133] A. Fölsing, *Albert Einstein: A Biography.,* E. Osers, Ed., Viking: New York, 1997.

[134] A. Einstein, "Remarks on Bertrand Russell's Theory of Knowledge" *In: The Philosophy of Bertrand Russell* The Library of Living Philosophers: Illinois: Evanston, 1944, pp. 278-292.

[135] A. Einstein, "Autobiographical Notes" *In: Albert Einstein. Philosopher-Scientist.* Illinois: Evanston, 1946, pp. 1-14.

[136] G. Holton, "Mach, Einstein and the Search for Reality", *Daedalus,* vol. 97, no. 2, pp. 636-673, 1968.

[137] A.I. Miller, "On Some Other Approaches to Electrodynamics in 1905." *In: Some Strangeness in the Proportion.* Addison-Wesley: N.Y., 1980, pp. 66-91.

[138] A. Einstein, "Motive des Forschens." *In: Zu Max Planks Sechszigstem Geburtstag. Ansprachen, gehalten am 26 April 1918 in der Deutschen Physikalischen Gesellchaft.* Muller: Karlsruhe, 1918, pp. 29-32.

[139] A. Einstein, "Physik und Realität", *J. Franklin Inst.*, vol. 221, pp. 121-164, 1936.

[140] T.S. Kuhn, *Black-Body Theory and Quantum Discontinuity, 1894-1912.* Oxford University Press: Oxford, New York, 1978.

[141] A. Einstein, "Eine neue Bestimmung der Moleküldimensionen". University of Zurich Dissertation, 1905. English translation by Anna Beck as "A New Determination of Molecular Dimensions". In: *The Collected papers of Albert Einstein. Vol. 2. The Swiss years: writings, 1900 – 1909.* Princeton, New Jersey: Princeton University Press, 1989, pp. 104-122. Published as an article in "Ann. Phys.", vol. 19, pp. 289-305, January 1906.

[142] A. Einstein, "Über die von der molekularkinetischen Theorie der Wärme geforderte Bewegung von in ruhenden Flüssigkeiten suspendierten Teilchen", *Ann. Phys.*, vol. 17, pp. 549-560, May 1905. English translation by Anna Beck as "On the Movement of Small Particles Suspended in Stationary Liquids Required by the Molecular-Kinetic Theory of Heat". In: *The Collected papers of Albert Einstein. Vol. 2. The Swiss years: writings, 1900 – 1909.* Princeton, New Jersey: Princeton University Press, 1989, pp. 123-134

[143] A. Einstein, "Eine Theorie der Grundlagen der Thermodynamik", Ann. Phys., vol. 11, pp. 170-187, January 1903. See also: A. Einstein, "Uber die thermodynamische Theorie der Potentialdifferenz zwischen Metallen und vollstandig dissozierten Losungen ihrer Salze und uber eine elektrische Methode zur Erforschung der Molekularkrafte", *Ann. Phys.*, vol. 8, pp. 798-814, 1902. [http://dx.doi.org/10.1002/andp.19023130806]

[144] A. Einstein, "Zur allgemeinen molekularen Theorie der Warme", *Ann. Phys.*, vol. 14, pp. 354-362, March 1904. English translation by Anna Beck in: *The Collected papers of Albert Einstein. Vol. 2. The Swiss years: writings, 1900 – 1909.* Princeton, New Jersey: Princeton University Press, 1989, pp. 68-77.

[145] M. Smoluchowski, "Zur kinetischen Theorie der Brownschen Molekularbewegung und der Suspensionen", *Ann. Phys.,* vol. 21, no. 14, pp. 756-780, 1906. [http://dx.doi.org/10.1002/andp.19063261405]

[146] A. Fulinski, "On Marian Smoluchowski's life and contribution to physics", *Acta Phys. Pol. B,* vol. 29, no. 6, pp. 1523-1537, 1998.

[147] W. Sutherland, "A dynamical theory of diffusion for non-electrolyte and the molecular mass of albumin", *Phil.Mag., S.6,* vol. 9, pp. 781-785, 1905. [http://dx.doi.org/10.1080/14786440509463331]

[148] A. Einstein, "Principé de relativité et ses consequencé dans la physique modern", *Archives des Sciences Physique et Naturelles,* vol. 29, pp. 125-144, 1910.

[149] J.D. Norton, "How Hume and Mach Helped Einstein Find Special Relativity." *In: Discourse on a New Method. Reinvigorating the Marriage of History and Philosophy of Science. Chicago and La Salle.* Open Court: IL, 2010, pp. 359-386.

[150] J. Stachel, "Introduction to Einstein: The Formative Years." *In: Einstein. The Formative Years, 1879-1909.* Birkhauser: Boston, 2000, pp. 1-21. [http://dx.doi.org/10.1007/978-1-4612-0311-7_1]

[151] G. Holton, "Mach, Einstein and the Search for Reality", *Daedalus,* vol. 97, no. 2, pp. 636-673, 1968.

[152] M. Slavov, "Empiricism and Relationism Intertwined: Hume and Einstein's STR", *Theoria,* vol. 31, no. 2, pp. 247-263, 2016.

[http://dx.doi.org/10.1387/theoria.14846]

[153] T. Ryckman, *The Reign of Relativity: Philosophy in Physics 1915-1925.* Oxford University Press: Oxford, 2005.
[http://dx.doi.org/10.1093/0195177177.001.0001]

[154] A. Pais, "Einstein and the Quantum Theory", *Rev. Mod. Phys.,* vol. 51, pp. 863-914, 1979.
[http://dx.doi.org/10.1103/RevModPhys.51.863]

[155] A. Einstein, "Ist die Trägheit eines Körpers von seinem Energienhalt abhängig?", *Ann. Phys.*, vol. 18, pp. 639-641, September 1905. English translation of Anna Beck as "Does the Inertia of a Body Depend on its Energy Content?" In: *The Collected papers of Albert Einstein. Vol. 2. The Swiss years: writings, 1900 – 1909.* Princeton, New Jersey: Princeton University Press, 1989, pp. 172 - 174.

[156] A. Einstein, "Über das Relativitätsprinzip und die aus demselben gezogenen Folgerungen", *Jahrbuch der Radioaktivität und Elektronik*: vol. 4, pp. 411-462, December 1907. English translation by Anna Beck in: *The Collected papers of Albert Einstein. Vol. 2. The Swiss years: writings, 1900 – 1909.* Princeton, New Jersey: Princeton University Press, 1989, pp. 252-311.

[157] A. Einstein, "Über die Entwicklung unserer Anschauungen über das Wesen und die Konstitution der Strahlung", *Phys. Z.*, vol. 10, pp. 817-25, September 1909. English translation by Anna Beck in: *The Collected papers of Albert Einstein. Vol. 2. The Swiss years: writings, 1900 – 1909.* Princeton, New Jersey: Princeton University Press, 1989, pp. 379 – 394.

CHAPTER 3

The Contrivance of General Theory of Relativity: Unification of Abraham, Einstein and Nordström's Hybrid Theoretical Models

Abstract: The aim of the chapter is to amend the received view on the general theory of relativity (GTR) genesis and advancement by taking into account common scientific practice of its functioning, the history of science data and certain philosophy of science arguments. The inter-theoretical aspect of the GTR genesis as an instance of perspicuous epistemological model of mature theory change that hinges upon 'old' theories encounter and interaction is scrutinized. I strengthen arguments in favour of the tenet that the dynamic creation of GTR had been continually governed by internal tensions between two research traditions, that of special relativity and Newton's gravity. The encounter of the traditions, their interpenetration and intertwinement engendered the rival programmes of relativistic theory of gravity construction: the programmes of Abraham and Nordström – on the one hand – and Einstein's programme – on the other. The encounter created the hybrid realm at first with an irregular set of theoretical models. The rival approaches of Einstein, Abraham and Nordström grew closer and eventually were transformed by Einstein into complementary 'mathematical' and 'physical' strategies of a general synthetic research programme that put forward the Entwurf and the GTR. Step by step, on eliminating the contradictions between the models contrived, the hybrid set was put into order by Einstein *via* the principle of equivalence. As in the STR case, Einstein was able to freely juxtapose Nordström, Abraham's and his own non-metric theoretical schemes without reducing one to the others. It is contended that the main reason for the GTR 'victory' over the rival programmes of Abraham and Nordström was a synthetic character of Einstein's programme. Einstein's programme did supersede the rival ones because it did ingeniously assimilate some ideas of the Nordström programme as well as some presuppositions of the programme of Abraham. Einstein had put forward as a basic synthetic principle the principle of equivalence that radically differed from that of rival approaches by its open, flexible and regulative character. As a result of reconciling and amalgamating the 'physical' and 'mathematical' approaches, embodied in Abraham, Einstein and Nordström's crossbred theoretical models , Einstein was able to explain the anomalous motion of Mercury.

Keywords: Abraham, Besso, Cartan, Christoffel, Einstein, Entwurf, Fokker, General relativity, Grossmann, Hybrid models, Laue, Levi-Civita, Maxwell, Mie, Minkowski, Neo-Kantian epistemology, Nonmetric theories, Nordström, Perihelion of Mercury, Planck, Principle of equivalence, Renn, Riemann, Special

Rinat Nugayev
All rights reserved-© 2018 Bentham Science Publishers

relativity, Stachel, Stepin, Unity of Nature, Zurich Notebook.

1. INTRODUCTION: WHAT IS WRONG WITH THE RECEIVED VIEW ON GENERAL THEORY OF RELATIVITY GENESIS

It is proverbial that Albert Einstein's titanic efforts to create the General Theory of Relativity (GTR) were accompanied by its rival flat versions conjured up by Gunnar Nordström, Max Abraham, Gustav Mie, *et al.* Already in 1912-1914 a Finnish mathematician Gunnar Nordström had advanced an ingenious scalar Lorentz covariant gravitation theory. In a paper submitted to '*Physikaliche Zeitschrift'* in October 1912, Nordström contended that Albert Einstein's bold conjecture that the speed of light c depends on the gravitational potential necessarily led to insurmountable difficulties disclosed in the Einstein-Abraham controversy. On the contrary, Nordström maintained that he had contrived a rival to Einstein's hypothesis which would "...leave c constant and still adapt the theory of gravitation to the relativity principle in such a way that gravitational and inertial masses are equal" (Nordström [1], 1126).

Likewise, in 1912 a Gottingen master of classical electrodynamics Max Abraham had invented a host of scalar and vector Lorentz covariant gravitation theories where light and gravitation had the same speed of propagation. In Abraham's scalar theories "c, the speed of light, depends on the gravitational potential. This hypothesis was first enunciated by Einstein (Ann. d. Physik, 35, 1911, p.898). Taking it as a starting point, I undertook to develop a theory of the gravitational field (M. Abraham, Physik. Zeitschrift 1912, p.1 and 311) which I then, in constructive competition with Mr. Einstein, gave a more satisfactory form" (Abraham [2], 793).

The papers of Abraham, Nordström, Mie *et al.* are still commonly taken as whimsical *delusions* that had been capable to stir up problem situations at best and to incite critical discussions around the GTR highlighting all its opulence and splendor. Eventually a received view on the genesis of the GTR was set up according to which, in the process of construction of the GTR, Einstein decidedly "rejected" Lorentz covariant scalar and vector theories of gravity. Indeed, §7 of Einstein & Grossmann's "*Entwurf*" pioneering metric paper bears the title "*Can the gravitational field be reduced to a scalar?*" In this paragraph, Einstein claims that any special-relativistic scalar theory of gravity necessarily violates energy conservation and therefore is physically untenable. Besides, Einstein repeatedly condemned Abraham's (1912a [3], 1912b [2]) new theory of gravitation, which was engendered from Einstein's idea of a variable speed of light as the gravitational potential.

> "Abraham [...] made some serious mistakes in reasoning so that things are probably incorrect. This is what happens when one operates formally, without thinking physically" (27 January 1912, Einstein to Heinrich Zannger; quoted from [4], 10).

Or, likewise,

> "Abraham's theory has been created out of thin air, *i.e.* out of nothing but considerations of mathematical beauty, and is completely untenable. How this intelligent man could let himself be carried away with such superficiality is beyond me" (Einstein to Michele Besso, quoted from [4], 10).

Note that at the same time Einstein continued to hold a high opinion of Abraham as a professional physicist. For instance, in 12 June 1912 letter to Ludwig Hopf he admitted that Abraham's theory was "truly superficial, contrary to his usual practice" (quoted from [5], p.10).

For either reasons, the alternative programmes were 'flatly abandoned'. However, some current history-of-science insights [5 - 7] prompt one to take the standard view with a considerable grain of salt.

To begin with, not everyone agreed with Einstein's aforementioned argument. Even after GTR proposal, Max von Laue wrote a paper on Nordström's theory, which he considered as physically consistent (although empirically wrong). Eventually, Einstein himself had to admit that Nordström's modified scalar theory was compatible with energy conservation, as well as with other reasonable constraints on possible gravitation theories [8]. Moreover, tentative study of the Einstein - Nordström correspondence surprisingly excavates that it was Albert Einstein himself who, before November 1915, and even *after* the creation of GR preliminary metric version – the notorious '*Entwurf*' (1913) – took active part in the invention of Nordström's scalar relativistic theories. Einstein was in close and constant contact with Nordström during the streak in which the Nordström theory was perfected. The theory actually advanced through an intensive exchange between Einstein and Nordström, with Einstein often generating the ideas crucial to the development of the theory. By and large the theory might more accurately be called the "Einstein- Nordström theory". It is not surprising that the theory became an account with kinematical effects very similar to general relativity. Moreover, gravitational fields would slacken the clocks and shorten the rods.

The next apparent example is Albert Einstein and Adrian D. Fokker's meticulous paper published in early 1914 that aimed at an "application of new mathematical methods, used in Einstein and Grossmann's paper, to Nordström's theory" ([9]). In the Introduction, Einstein and Fokker promised to exhibit that "...one arrives at Nordström's theory instead of the Einstein-Grossmann theory if one makes the single assumption that it is possible to choose the preferred reference systems in such a way that the principle of the constancy of light obtains" (quoted from [5],p.60). However, there subsided a certain irony in the paper of Einstein and Fokker, the irony that their readers would reveal within subsequent two years .While the use of preferred coordinate systems was posited against Nordström's subtle theory, Einstein's own 'Entwurf' theory was not itself generally covariant [5].

Moreover, in the same paper in early 1914 the significant connections between Nordström's exquisite theory and conformally flat spacetimes were revealed. Therefore, it comes as no surprise that it was within Nordström's 1912 theory where the gravitational field equation $R = \varkappa T$ ($\varkappa$ = const) was first derived, with R being fully contracted Riemann-Christoffel tensor and T the trace of the stress-energy tensor. This field equation is a blatant harbinger of Einstein's stupendous equations presented to Preussiche Akademie der Wissenschaften on November 25, 1915 [10].

As a consequence of the successful reformulation of Nordström's theory in generally-covariant terms, it was only natural to search for an analogous reformulation also of the '*Entwurf*' theory. That such a venture made sense was obviously the conclusion which Einstein and Fokker drew at the end of their thought-provoking paper:

> "Finally, the role that the Riemann-Christoffel differential tensor plays in the present investigation suggests that this tensor may also open the way for a derivation of the Einstein-Grossmann equations that is independent of physical assumptions. The proof of the existence or nonexistence of such a connection would represent an important theoretical advance" [9].

Einstein made clear his preference for Nordström's ingenious theory over other rivals in September 1913 epoch-making presentation of the '*Entwurf*' theory (see section 3 for details) to the 85th Congress of the German Natural Scientists and Physicians in Vienna. His single critical remark consisted in that the Nordström theory was incompatible with Mach's principle – a vice that could turn out a virtue to a Naturforscher biased against metaphysical castles in the air. In 1921, none other than Wolfgang Pauli christened this theory an 'empirical blunder' since it had not predicted any deflection of a light ray by a gravitational field and had not explained the anomalous motion of Mercury.

> "The theory solves in a logically *quite unexceptionable* way the problem sketched out above, of how to bring the Poisson equation and the equation of motion of a particle into a Lorentz-covariant form. Also, the energy-momentum law and the theorem of the equality of inertial and gravitational mass are satisfied. If, in spite of this, Nordström's theory is not acceptable, this is due, in the first place, to the fact that it does not satisfy the principle of general relativity (or at least *not in a simple and natural way*). Secondly, it is in contradiction with experiment: it does not predict the bending of light rays and gives the displacement of the perihelion of Mercury with the wrong sign. (It is in agreement with Einstein's theory with regard to the red shift" ([11], p.144)

Yet there had been no eclipse expeditions in 1913 and Einstein's own '*Entwurf*' (due to Michele Besso's assiduous calculations) turned out to be incapable to explain the anomalous motion of Mercury.

On the contrary, in 1912 G. Pavani thoroughly calculated the perihelion shift of Mercury according to Abraham's theory, obtaining value of 14'' 52, that is approximately one third of the observed one. Thus, Abraham's vector theory made a more accurate prediction than the '*Entwurf*'. If one takes into account that Einstein's discovery (due to Michele Besso's diligent calculations) of the failure of the '*Entwurf*' theory to yield the correct perihelion shift of Mercury was made as early as the summer 1913, one can conclude that Abraham's advancement was a vigorous spur that prompted Einstein to construct GTR out of the '*Entwurf*' .

Furthermore, the consequences of the '*Entwurf*' and the GTR coincide with the consequences of the theories of Nordström and Abraham for a number of important cases in certain wholesome approximations. For instance, the '*Entwurf*' is reduced to a theory with a four-vector field potential that is formally analogous to maxwellian electrodynamics in suitable weak - field approximation. Moreover, special theory of relativity turns out to be an inescapable intermediary step in the transition from GTR to Newton's theory [12]. But this transition is grounded on the supposition, for weak and stationary gravitational fields, that the gravitational field is described by a scalar in flat (Minkowski) space- time, *i.e.* on the reduction to scalar Nordström's theory [6].

Besides, the so-called 'linear approximation' in GTR ([13]), still in common use to account for propagation and detection of gravitational waves, especially in 2016-17, presupposes the transition to such a theory of gravitation in which the gravitational wave, in full analogy with classical electrodynamics, is described by a vector in flat space- time, *i.e.* the transition to vector theory of Abraham [14]. Abraham's ambitious claim that his theory contained Einstein's as a limiting case was rebutted by Einstein, though. But, nevertheless, in both cases the relations between the GTR and the theories of Nordström and Abraham strongly resemble the common pattern of classical electrodynamics where the general potential is represented by many-component object such as a vector or a tensor, which, in the special case of a static field, reduces to a single-component object.

Moreover, some new light on the interrelations between Einstein, Abraham and Nordstrom's theories was brought by the so-called '*Zurich Notebook*' that was employed by Einstein while constructing his '*Entwurf*' theory, where the entries begin in mid-1912 and end in early 1913. As the '*Zurich Notebook*' indicates, the pivotal requirement constraining the choice of gravitational field equations for Einstein in 1912-1913 was that they reduce to the field equation of Newtonian theory in the case of weak static fields.

However, John Stachel [15] shrewdly commented that Einstein *actually* compared the field equations of the '*Entwurf*' with Newtonian theory in the case of weak, static fields by dint of presuming an *excessively restrictive* form for the metric representing such fields. Stachel elicited that Einstein was 'handicapped' in the comparison of his theory to Newtonian theory of gravitation by the mathematical apparatus at hand. For instance, he lacked the subtle concept of an 'affine connection' that was to appear only ten years later in the writings of Elie Cartan *et al.* In that case Einstein expected that the metric tensor for the case of weak static fields would have only one variable component, which would behave like the Newtonian gravitational field. Since the remaining components were *assumed* to be constant, the spatial geometry would just be ordinary Euclidean one. Yet this reasonable expectation obviously contradicted the GTR generally-covariant field equations Einstein finally adopted in 1915 and could by itself preclude their adoption. So he had to impose notorious '*harmonic coordinates*' restrictions. Withal Einstein revealed that the coordinate restrictions he stubbornly applied were closely associated with energy-momentum conservation in the weak-field limit of the 'Entwurf'. This circumstance obviously suggested to him that the full theory could only be made compatible with the principle of energy-momentum conservation at the expense of restricting its covariance.

Thus, one has to recapitulate that Einstein did not bother himself by taking into account the peculiarities of Riemann geometry and *actually*, drawing upon physical considerations, took $g_{\mu\nu}$ as a common tensor in flat space- time, eliminating the corresponding terms according to physical sense considerations. It's no wonder if one takes into account Einstein's constant preoccupation with the analogies with maxwellian electrodynamics. But namely this premise is the crux of Abraham's and Nordström's scalar and vector nonmetric theories. It should be added that the absence of generally covariant equations to compare the Entwurf equations with gave this theory a contingent character; M. Abraham and another sagacious critic, G. Mie, punctuated this problematic absence in their publications (see [4] for details).

It is no wonder that the startling results that Abraham obtained in the course of his research, such as the very possibility and some essential properties of gravitational waves, remain to this day a standard for a relativistic theory of gravity [16]. Moreover, Abraham, already in March 1912, was one of the first researchers to hit upon a singularity in a field theory of gravitation and to calculate what was later called 'the Schwarzchild radius' of a nonrotating black hole [17]. It is well-known that only *after* Abraham's first publication did Einstein also turn to the problem of gravitation after a period of 1908-1911 vehemently devoted to quanta.

Furthermore, Einstein's approach to creation of a relativistic theory of gravity, - on the one hand, - and Nordström and Abraham's approaches, - on the other hand, - where in many respects ***complementary***. For instance, Einstein's theory of static gravitational field [18] was incited by substantially physical considerations based on the equivalence principle, while Abraham's theories started from mathematical considerations related to Minkowski formalism. Yet the diligent investigation of the '*Zurich Notebook*' makes it possible to bring the following new light on Einstein's thought laboratory.

> "In the course of Einstein's work documented in the Zurich Notebook, the two approaches [*i.e.* of Einstein and Abraham] grew closer and turned into *complementary strategies* of a more or less systematic research program. In this research the two approaches were distinguished mainly by the sequence in which the building blocks of the theory [*i.e.* the Poisson's equation, the covariance principle and the restrictive conditions for weak and static fields] come into play" ([6],125; my italics).

Though in his younger years Einstein had had a 'lukewarm' (van Dongen's apt term) attitude towards mathematics, the '*Zurich Notebook*' unfolded that, in the course of the '*Entwurf*' construction, he was mostly working on a *dual* strategy that deftly ***embraced*** both physical and mathematical approaches to finding the appropriate field equations. This strategy

> "was an iterative process that began with trying out one approach, and then checking results with demands or results of the other approach" ([4],12).

Hence Einstein's famous "physical" and "mathematical" strategies, the vacillation between whom brought to the '*Entwurf*' and the GTR construction, were entrenched in two research traditions encountered circa 1907.

On the other hand, for a long time it was a commonplace that creation of the GTR embodies an exceptional case in the history of physics since it was put forward by a single person – Albert Einstein. Owing to this reason, the theory grounded on extremely narrow empirical base (three critical effects of GTR) managed to provide an avalanche of empirical data referring to relativistic astrophysics and cosmology. The both hallmarks are habitually explained *via* a romantic image of a Lone Genius Grasping the Essence of Things in the Bursts of Divine Inspiration.

Nevertheless, in spite of overwhelming importance of Einstein's impact, this explanation seems untenable in the light of current post-Kuhnian philosophy and sociology of science. A transition from the 'old' paradigm to a 'new' one can be taken as such only if it involves the majority of the members of the scientific community. And only that explanation for the origins of the paradigm change in science gets the obvious advantage over the other accounts that contemplates the behavior of the *majority* of the scientists as lucid, tentative and sagacious (as rational).

As it is elucidated in the monumental study,

> "both the peculiar emergence and the *remarkable stability* of Einstein's theory of gravitation with regard to the further development of physics and astronomy becomes plausible only if the genesis of general relativity is understood, not as a fortunate anticipation of future observational discoveries, but as a *transformation of pre-existing knowledge*" ([19], 23; my italics).

All in all, the dynamics of the (relativistic) theory of gravity was predominantly governed by significant *internal tensions*, profound contradictions within the knowledge system rather than by new empirical knowledge, which played only a subordinate role at best [20]. The anomalous perihelion of Mercury remained a commonly used touchstone for all theories of gravitation for more than a half century before the contrivance of GTR. The famous bending of light in a gravitational field could simply be deduced from the apparent observation that, in an accelerated frame of reference, light rays must be curved as a consequence of the superposition of the motion of the observer and of the light. Successful, or decisive red shift experiments were performed only in 1960's.

In his 1949 autobiography Einstein famously wrote:

> "I have learned … from the theory of gravitation: no collection of empirical facts however comprehensive can ever lead to the formulation of such complicated equations [of GTR]. A theory can be tested by experience, but there is no way from experience to the construction of a theory. Equations of such complexity as the equations of the gravitational field can be found only through the discovery of a logically simple mathematical condition that determines the equations completely or almost completely. Once one has obtained those sufficiently strong formal conditions, one requires only little knowledge of facts for setting up a theory; in the case of the equations of gravitation it is the four-dimensionality and the symmetric tensor as expression for the structure of space which, together with the invariance concerning the continuous transformation-group, determine the equations almost completely" (quoted from [4], p.32).

All the above mentioned conspicuous hallmarks of the GTR genesis, advancement and functioning, and preponderantly the current modus operandi, ***the common practice of its implementation*** bolster the following conclusions.

a. The relations between the GTR and its ingenious rivals were far more complicated in 1907-1915 than it may seem from the notorious 'truth-falsity' dilemma, so that one can contemplate the interlacement and *interpenetration* of rival 'paradigms' into each other.
b. The rival approaches of Einstein, Abraham and Nordström grew closer and eventually were ingeniously turned out by Einstein into complementary 'mathematical' and 'physical' strategies of a systematic *synthetic* research programme that engendered the '*Entwurf*' and the GTR. Hence it's no wonder that Einstein's GTR was better than its abovementioned rivals if only for the reason that it *encompassed them all* in significantly modified forms. (Just as

the GTR embraces Newton's theory of gravitation and the STR, or just as the maxwellian electrodynamics encompasses the partial theoretical schemes of Coulomb, Ampére, Biot & Savare *et al.*).

c. Einstein could complete the reconciliation of the knowledge on gravitation and inertia (represented by classical mechanics) and the knowledge on the structure of space and time (embodied by STR) *via* the 'Entwurf – GTR' transition only. As a result, he was able to explain the anomalous motion of Mercury.

Hence **the aim of the chapter** is to modify the standard view on the GTR genesis and advancement by digging still deeper into the inter-theoretical context of the GTR genesis , that is, by taking into account the above mentioned history of science data, philosophical (and sociological) arguments and modern common scientific practice of its functioning. My **main idea** consists in that the basic reason for the GTR victory over the rival programmes of Abraham and Nordström lied in a *synthetic* character of the Einstein research programme. Einstein's research programme did supersede the rival ones because it did ingeniously assimilate some premises of the Nordström programme as well as some presuppositions of the programme of Abraham. Einsteinian programme's victory over its ingenious rivals became possible since Einstein had put forward as a basic synthetic principle the principle of equivalence that substantially differed from that of rival accounts by its open, flexible and regulative, 'heuristic' character.

In the **second section** of the chapter an epistemological model that fits the accomplishments of current philosophy of science and deals with mature theory dynamics and structure is exhibited in the gravitational context. The model is a crux of the present study asserting that the history of physics does not advance through the creation *ex nihilo* of new paradigms, but rather through long-term processes of tentative readjustment, reconciliation, interpenetration and intertwinement of 'old' research traditions.

In the **third section** of the chapter the initial stage of GTR creation (1907-1912) is scrutinized. The gist of the section is an assertion that the invention of relativistic theory of gravity had commenced with the crossbred object construction in Einstein's 1907 paper, *i.e.* with the introduction of mass-energy relation into the theory of gravity. The crossbred object entry – the introduction of inertial and simultaneously gravitational mass – turned out to be crucial to a penetration of STR methods into the Newtonian theory of gravity and to a reverse penetration of Newtonian gravity methods into the STR. As a result, the both theories were radically restructured from within and the corresponding changes in both of them were set up. The changes were epitomized in the peculiar sequences of crossbred models, the byproducts of the transformation performed.

i. On the one hand, an inevitable consequence of the relentless penetration of STR into Newtonian gravitation theory turned out to be Nordström's and Abraham's research programmes.
ii. On the other hand, no less inevitable, due to the equivalence principle, was the Newtonian theory penetration into the STR. It led to the sequence of Einstein's writings on the generalization of relativity principle and to spreading the principle not only on inertial systems of reference, but on the various accelerated systems as well. Though the most valuable result of the hybrid theories of Nordström and Abraham consisted in that both of them contained some *extremely promising hints* on how the global theory could be created. As a result, the GTR emerged as a promising 'synthetic approach' unifying all the positive achievements of the other alternative accounts. Hence the climax of the strake was Einstein's proposal and apprehension of the equivalence principle that became one of the firm GTR heuristic foundations.

The **fourth section** of the chapter (1912-1913) scrutinizes the '*Entwurf*' contrivance. This metric theory had sprung out not as 'phoenix from the ashes' but *via* the meticulous synthesis of Abraham and Nordström's theoretical schemes, as well as from the preliminary nonmetric theoretical schemes of Einstein. The staple was the metric tensor construction owing to equivalence principle and Nordström, Laue and Planck's startling results. And *it is namely the fact that the basic model of the Entwurf was constructed due to subtle unification of Nordström, Abraham and Einstein's (obtained before 1913) theoretical schemes that can explain the reasons for Einstein's programme victory over its rivals.* For any theory unification becomes a 'true' one only in the case when the corresponding research traditions encountered are *successfully* reconciled, *i.e.* when their conjunction leads to novel firm, concrete and empirically verifiable results (the perihelion shift of Mercury being a blatant example).

The **fifth section** grapples with 1913-1915 transition from the '*Entwurf*' to the fool-blooded GTR exposed by Einstein in the lecture to Berlin Academy on November 25, 1915. It is demonstrated that the main thrust in passing from the '*Entwurf*' to the GR came from Einstein's 'dual strategy' inducing him to *reconcile* physical and mathematical approaches. And only *after* he efficiently reconciled them. *i.e.* only after he realized the general covariance of the gravitational field equations could he turn to successful explanation of the Mercury perihelion. Hence the decisive impetus in constructing the General Relativity came from the attempts to reconcile physical and mathematical approaches, embodied in Abraham, Nordström and Einstein's crossbred theoretical models. Trustworthy explanation of Mercury perihelion motion appeared to be a by-product of strenuous reconciliation efforts.

Riemannian geometry was deployed as a "neutral language" to embrace and co-ordinate the theoretical languages encountered and to provide successful explanation of Mercury perihelion shift.

2. A LUCID EPISTEMOLOGICAL MATURE THEORY CHANGE MODEL

As is mentioned in the Introduction, the dynamics of the theory of gravity was predominantly governed by *internal tensions*, profound contradictions within the knowledge system rather than by new empirical knowledge, which played only a subordinate role at best. In this section, a corresponding epistemological model dealing with mature theory dynamics and structure is posited that is apt to reconstruct epistemologically an encounter and interaction of different theoretical research traditions.

The current philosophy of science discourse on scientific revolutions allows one to elucidate the views on the structure and functioning of scientific theories, – on the one hand, - and to construct sufficiently sweeping and exact theory change epistemological models, – on the other. In particular, according to one of the current models [21], a scientific revolution is triggered by encounters of some entrenched "old" paradigms, scientific research programmes or research traditions that cannot be reconciled in a common way, *i.e.* by reducing of one of them to another. The way out of the quandary is to work out such a global theory that successfully encompasses all the theories involved (in significantly modified forms, as a rule). The global theory is aimed at eliminating the tensions, smoothing away dissensions between 'incommensurable' paradigms involved.

Just to recapitulate Werner Heisenberg's stupendous "*Physik und Philosophie*" (Frankfurt am Main, 1959):

> "probably, as a kind of general supposition, it can be said that those directions in the history of human thought appeared to be most fruitful, where different ways of thinking had encountered. These ways of thinking are deeply rooted in different spheres of human culture, or in different times, in different cultural milieu, or in different religious traditions. When they really meet with each other, when they correspond to each other so that an *interaction between them takes place*, one hopes that new and interesting discoveries will follow" (my italics).

In the course of global theory invention an indispensable preliminary stage shows itself in the contrivance of a series of hybrid ('crossbred') theories. The latter are persistently set up to the climax when such a hybrid model is contrived that is able to outline the fruitful way of the global model creation through the generalization of models that belong to the lower level of mature theories. According to the aforementioned lucid epistemological model, scientific revolutions or ***radical breakthroughs in science were not due to ingenious invention of new paradigms or the fanciful creation of new ideas ex nihilo, but rather to the long-term processes of the reconciliation, interpenetration and intertwinement of 'old' research traditions preceding such breaks.***

It is a commonplace that no judicious epistemological model of scientific revolutions can be established without preliminary elucidating the structure of mature theories. Yet what I want to underscore is that a mature scientific theory of XIX and XX centuries physics encompasses not a single model or a bundle of models. It embraces a *bundle of groups of models* that are related to one another in rather subtle ways. A mature theory is so structured that the multitude of its models is disseminated over at least three interconnected levels [22].

1. The level of the ***basic*** model (or 'the Fundamental Theoretical Scheme').
2. The level of the *subordinated* models (or 'the Partial Theoretical Schemes') constructed from the fundamental one according to certain (tacit) rules.
3. The level of the '*Empirical Schemes*' that can be approached through the level of partial theoretical schemes.

For instance, the relations between the basic objects of classical (Newtonian) mechanics are highlighted by Newton's laws. The derivative objects of classical mechanics are 'an absolutely rigid body', 'a central field', 'a harmonic oscillator', and so on. The relations between them are fixed by certain laws of classical mechanics: that is, by 'the laws of rigid rotation', 'movement in the central field', 'laws of small oscillations', and so on.

Likewise, the electric field at a point **E**, the magnetic field at a point **H**, the current density **J** and the charge density ρ are the basic theoretical objects of classical (Maxwellian) electrodynamics. Maxwell's equations elucidate the relationships between them.

The set of basic objects of a mature physical theory's constitutes the *basis*, *i.e.* the subsystem of theoretical objects. All the basic theoretical objects are blatant idealizations and cannot exist as real bodies (like tables and chairs or rods and clocks). For instance, "the material point" is taken as a body free of dimensions. As for the other basic objects of classical mechanics, it is assumed that 'an inertial system of reference' can be successfully isolated from external influence.

The derivative subsystems are *subordinated* (Vyacheslav Stepin) to the basic one, but are (relatively) independent of each other, referring to different aspects of the same domain of validity. Each subsystem is properly characterized by its own set of notions and mathematical apparatus that belong to a special part (section) of the mature theory.

For instance, Newtonian mechanics consists of several independent sections: 'small-oscillations mechanics', 'mechanics of rigid body rotations', 'mechanics of movement in a central field', and so on. Each section is characterized by its own subsystem of derivative objects. Each subsystem constitutes a subtle model of a particular type of mechanical motion (the model of small oscillations, the model of rigid rotations, and so on). Relations between the elements of the subsystem are fixed by particular laws of Newtonian mechanics.

All in all, the complicated relations between a subsystem of basic theoretical objects and a subsystem of derivative ones can be elucidated as follows. Any derivative subsystem is constructed from the basis by a process of *reduction*. That is, a mature theory is uncoiled not by formal (logical, mathematical) means only, but also through subtle *gedanken experiments* with abstract theoretical objects. The reduction is put into effect by taking into account the peculiarities of the empirically fixed domain of validity. This domain can be 'seen through' a cognitive lens of an ideal model, formed by correlations of basic objects. According to the peculiarities of each concrete experimental situation, various constraints may be imposed on the system of basic theoretical objects. This enables us to define the system, transforming it into a subsystem of derivative objects. The fundamental equations are then applied to the subsystems of derivative objects. In accordance with the system features, they are transformed into the partial laws. The informal nature of such manipulations converts such an inference into a special problem solving operation (a 'puzzle' in Thomas S. Kuhn's vocabulary). The solutions of such conundrums are included in a theory at its origin (Ptolemy's "*Almagest*", Newton's "*Principia*", Maxwell's "*Tractatus*", Einstein and Grossmann's "*Entwurf*" and so on). To a theoretician bothered by applying a theory, they serve as a pattern for subsequent activity. Each problem is solved in accordance with primary '*paradigms*' (Thomas S. Kuhn).

In Newtonian mechanics, the paradigm examples include 'derivations' from Newton's laws: the law of small-oscillations, the law of movement in a central field law, the law of rigid body rotations and so on. In Maxwellian electrodynamics the paradigm examples include the laws of Biot and Savart, Coulomb, Ampére, Faraday *et al.*, derived from the Maxwell equations.

In general theory of relativity, the host of paradigm examples includes the derivation of Newton's theory of gravity from the Einstein equations in the "*weak field approximation*". In the case of weak gravitational field such a system of reference is commonly chosen in which all the metric tensor components slightly differ from their Minkowski values: g = η + h [12]. The further demand to ignore the squares and the other multiples of h is necessary for the transition to Newton's theory of gravitation. But all these means nothing else that index rising operation should be carried out by η – the metric tensor of **flat** space-time. As a result, in the so-called " weak field approximation" the gravitational field equations take the form of usual wave equation in flat space time for (Nordström's) scalar potential φ . Thus the basic theoretical object of Nordström's nonmetric scalar theory turns out to be constructed from the GTR basis. "This is quite natural since the weak field is considered as a tensor in flat space-time" ([16], 56) and is described by an equation

$$h_{ik} = - \eta_{ik} \, 2\varphi/c^2.$$

Meticulous **construction** of the derivative objects from the basic ones enables one to compare theoretical knowledge with experimental results, to explain and predict the results of real experiments. To this end, an empirical equation - an intermediate relation - is derived from the partial law. Through this equation the special constructs are inserted. Contrary to abstract objects, the newly born constructs are no longer idealizations and can be compared with real experimental devices now. These constructs are called *empirical objects*, and their systems - special representations of empirical situations - are christened '*empirical schemes*'. Empirical objects are not equivalent to real bodies. An empirical object cannot be compared with a single body with which an experimentalist operates, but only with a *class* of such objects. Consequently, an empirical scheme corresponds not to a concrete experimental situation, but to a *type* of such situations (V.S. Stepin). For instance, the empirical scheme of the Biot and Savare experiment with a magnetic needle and a conductor refers to any experiment with any current in the conductor and any sufficiently small magnetic needle.

A mature theory becomes an **established** one when the links between all the three levels of the organization are vigorously installed that makes possible to use the mature theory as an effective vehicle for making successful predictions. All the bonds between all the three levels of an established mature theory should be fairly robust ones. This rigidity allows us to connect a prediction referring to the upper level with all the levels of a mature theory. Hence it allows one to construct an experimental device to check the prediction. A new result, obtained in the advancement of mathematical apparatus, immediately influences all the levels of a mature theory. Hence a mature theory can predict, and the predictions can be verified. A mature theory obtains the status of an established one when at least some of its predictions are exhibited to be successful. It demonstrates that the system of basic objects is complete, and the links between all the three levels are made robust.

Due to complicated structure of a mature theory, creation of the global theory turns out to be a slow, adamant and consequent ascent from the lower levels up to the top ones. Any transition from lower level to the upper one is impossible until the construction of all the lower-level models is perfected. Withal the lower models (that served at scaffolding the upper ones) are not eliminated. They can be found not only in history-of-science archives. They can be transpired in real common practice of theories' functioning (in implicit forms, as a rule).That is why I think that the basic models of Nordström and Abraham's theories constitute partial theoretical schemes of the GTR.

3. CONSTRUCTION OF THE HYBRID MODELS' SEQUENCES *VIA* THE EQUIVALENCE PRINCIPLE

The advent of the special theory of relativity (STR) and the blatant and profound incompatibility between Newton's theory of gravitation and the STR perplexed Einstein and his contemporaries with the task of contriving an adequate relativistic theory of gravitation. Apparent *contradictions* between these theories consisted first and foremost in the fact that according to Newton's theory the velocity of gravitational interaction was infinite. On the other hand, STR prohibits the signals travelling faster than light. Undisguised disparity between the concepts of 'action at a distance' and 'instantaneous action' was revealed just after the construction of maxwellian electrodynamics. It was James Clerk Maxwell himself who tried to invent the first vector theory of gravity. Alas, he was forced to leave the efforts soon due to the problem of negative energy of gravitational wave. STR creation only exacerbated the problem [23, 24].

In his pre-eminent1906 paper "*Sur La Dynamique de l'Electron*", scrutinized in the preceding chapter, Henri Poincaré had revealed two constraints a relativistic law of gravitation should obey.

1. The new force could no longer depend only on the masses of the gravitating bodies and the distances between them. First of all the new force should depend on their velocities.
2. Gravitational action should not be taken as instantaneous, but should propagate with certain finite velocity.

However, it was Einstein's 1907 eminent review "*On the Relativity Principle and the Conclusions Drawn from it*", published in Johannes Stark's "*Jahrbuch der*

Radioaktivität und Elektronik", that laid the true conceptual foundations for all the relativistic theory of gravity.

> "The most important result of the fourth part is that concerning the inertial mass of the energy. This result suggests the question *whether energy also possesses heavy (gravitational) mass*. A further question suggesting itself is whether the principle of relativity is limited to nonaccelerated moving systems. In order not to leave this question totally undiscussed, I added to the present paper a fifth part that contains a novel consideration, based on the principle of relativity, on acceleration and gravitation" (Einstein 1907 [25], 254-255;my italics).

And in the fifth part of his epoch-making 1907 paper Einstein boldly formulated first his "*principle of equivalence*". As he later recalled, when he had furnished the 1907 review article for publication, he had tried to modify Newton's theory of gravitation so as to reconcile it with the STR.

In 1933 lecture, in an outline of his pathway to GTR, Einstein famously recalled:

> "Like most writers at the time, I tried to frame a field-law for gravitation, since it was no longer possible, at least in any natural way, to introduce direct action at a distance owing to the abolition of absolute simultaneity.
>
> The simplest way, of course, to retain the Laplacian scalar potential of gravity, and to complete the equation of Poisson in an obvious way by a term differentiated with respect to time in such a way that the special theory of relativity was satisfied. The law of motion of the mass point in a gravitational field had also to be adapted to the special theory of relativity" (quoted from [5], p.5).

Thus, the corresponding attempts had demonstrated that it was possible, but Einstein had relinquished them soon since they were grounded on physically inacceptable hypotheses.

> "At this point, there occurred to me the happiest thought in my life [der glücklichste Gedanke meines Lebens]. ***Just as in the case with***

> ***the electric field*** produced by electromagnetic induction, the gravitational field has similarly only a relative existence. For if one considers an observer in free fall, *e.g.* from the roof of a house, there exists for him during this fall no gravitational field – at least not in his immediate vicinity. Indeed, if the observer drops some bodies, then these remain relative to him in a state of rest or in uniform motion, independent of their particular chemical or physical nature" (quoted from [26], 178 ; my italics).

Since the importance of the Equivalence Principle for the creation of GTR cannot be overestimated and due to abiding discussions on its 'true content', we have to resort to all the piece of the 1907 paper where the principle had been first formulated.

> "We consider two systems $\sum_1$ and $\sum_2$ in motion. Let $\sum_1$ be accelerated in the direction of its X-axis, and let γ be the (temporally constant) magnitude of that acceleration. $\sum_2$ shall be at rest, but it shall be located in a homogeneous gravitational field that imparts to all objects an acceleration $-\gamma$ in the direction of the X - axis.
>
> As far as we know, the physical laws with respect to $\sum_1$ do not differ from those with respect to $\sum_2$; this is based on the fact that all bodies are equally accelerated in the gravitational field. At our present state of experience we have thus no reason to assume that the systems $\sum_1$ and $\sum_2$ differ from each other in any respect, and in the discussion that follows, we shall therefore ***assume*** the complete physical equivalence of a gravitational field and a corresponding acceleration of the reference system.
>
> This assumption extends the principle of relativity to the uniformly accelerated translational motion of the reference system. The ***heuristic value*** of this assumption rests on the fact that it permits the replacement of a homogeneous gravitational field by a uniformly accelerated reference system, the latter case being to some extent accessible to theoretical treatment" (Einstein 1907 [25], 302; my italics).

Note that Einstein was first and foremost interested not in disguising the ontological, metaphysical content of his principle that could enable him to elevate the tenet up to the status of an Ultimate Law of Nature. The latter would be valid everywhere with any degree of validity being contemplated by a kind of Super Reason trying to grasp the essences of the things and events. (For it is well-known [5] that in 1907 Einstein was unaware of Eotvös's startling experimental results stating the equality of inertial and gravitational mass. Furthermore, in 1951 Papapetrou [27] had ascertained that in the GTR a rotating body falls differently from a non-rotating one). On the contrary, recalling the history of invention of the Equivalence Principle, Einstein appealed not to, say, metaphysical systems of Aristotle or Hegel that strongly encouraged grasping the "essences" of gravitational phenomena but *to his own experience of STR contrivance* [26]. It seems evidently rational for Einstein of 1907 to repeat once more the steps that had led to success just in 1905.

Thus, both in STR and GTR cases he was looking for the *heuristical* aspects of the principle [28]. In gravity purview he aimed at comprehending gravitational and inertial phenomena from a single point of view. As Michele Janssen puts it,

> "While the slide into general covariance turns the relativity of non-uniform motion of space-time coordinate systems into a feature general relativity shares with older theories, it does not so trivialize the relativity of the gravitational field. Even in generally covariant reformulations of these older theories, there will be an inertial field and a gravitational field existing side by side. The *unification of these two fields into one inertia-gravitational field* that splits differently into inertial and gravitational components in different coordinate systems is one of Einstein's central achievements with general relativity" ([29], 162; my italics).

In my view, it was consequent implication of the Equivalence Principle that promised to construct a sequence of hybrid models unifying STR and Newton's theory of gravity. The Principle of Equivalence was for Einstein not so much a Law of Nature as a ***pattern***, a 'paradigm' for construction of gravitation theories. In a sense, the pivotal hallmark of Einstein's scientific style, of his actual *modus operandi* was his fine anti dogmatism. As in the STR case, Einstein was able to freely juxtapose Nordström, Abraham's and his own non-metric theoretical schemes without reducing one to the others (see the next sections for details).

In particular, it enabled the investigation of special cases of the gravitational field by means of the study of accelerated motions. Thus, until 1911 Einstein had committed himself mainly to investigating, by means of the Equivalence Principle, the effects and conceptual changes characterizing a new theory of gravitation, evidently without seriously attempting its contrivance. Only in early

1912 was he challenged by the provoking publication of Max Abraham to elaborate such a theory, at least for the special case of a static gravitational field [30].

Moreover, in September 1913 Einstein presented an eminent lecture at the 85th Congress of the German Natural Scientists and Physicians in Vienna that was quickly published in December issue of "*Physikalische Zeitschrift*" under the heading "*On the present state of the problem of gravitation*". In this lecture Einstein disclosed his preference for Nordström's theory over other gravitation theories. He maintained that Nordström's later version of his gravitation theory was the only competitor to the '*Entwurf*' theory satisfying four important constraints that could be imposed on any reasonable theory of gravitation.

1. Satisfaction of the laws of energy and momentum conservation.
2. The Equivalence Principle.
3. Validity of STR.
4. The observable laws of nature do not depend on the absolute magnitude of the gravitational potentials.

Note that Einstein stressed the *heuristic* value of almost all the requirements admitting that "the postulates 2-4 resemble a ***scientific profession of faith*** more than a firm foundation" (Einstein 1913 [31]).

On the other hand, the second substantial component of Einstein's heuristic – "the Lorentz model of a field theory" (Renn and Sauer) – enabled Einstein to take Newtonian gravitation and inertia as special cases of a more general interaction. For the case of uniform acceleration he could directly identify inertial effects with a scalar Newtonian gravitational field and he expected that he would be able to do the same for more general cases by generalizing the notion of gravitational field. A paradigm for the generalizations was evidently provided by relativistic electrodynamics. It was Einstein's own special theory of relativity (STR) that 'unified' electricity and magnetism through treating electric field **E** and magnetic field **B** as different aspects of electromagnetic field tensor $F_{\mu\nu}$. Accordingly, for Einstein the most important achievement of GTR was not 'geometrization of gravity' but "unification of gravity and inertia" *via* the metric tensor $g_{\mu\nu}$.

In one of his writings [32] Einstein in 1912 even punctuated the 'equality of essence' [Wessengleichheit] of inertial and gravitational mass. Between 1907 and 1911 he deftly used the Equivalence Principle to derive several important consequences of his yet to be formulated relativistic theory of gravitation.

Note that in the case considered Einstein follows the path of the STR genesis. Indeed, the invention of new theory begins with the construction of the crossbred object, *i.e.* with the introduction of the mass-energy relation into theory of gravity. One of the important STR consequences is the tenet of equivalence of mass and energy. Yet, according to Einstein,

> "this result suggests the question whether energy also possesses heavy (gravitational) mass. A further question suggesting itself is whether the principle of relativity is limited to nonaccelerated moving systems" (Einstein 1907 [25], 254).

From the outset Einstein was seeking such a theory of gravitation that was to embrace both the knowledge on gravitation and inertia represented by the classical mechanics and the knowledge on the structure of space-time embodied by the STR. However, introduction of the crossbred object – the insertion of inertial and simultaneously gravitational mass – leads to gradual penetration of STR methods into Newtonian theory of gravity and to reverse and no less inevitable penetration of methods of Newtonian theory of gravitation into the STR. As a result, the both theories were radically restructured from within and the corresponding changes in both of them were set up. The changes were epitomized in the peculiar sequences of crossbred models, the byproducts of the transformation performed.

On the one hand, an inevitable consequence of STR penetration into Newtonian theory of gravity turned out to be Nordström and Abraham's scientific research programmes. On the other hand, no less inevitable, owing to the equivalence principle, was Newtonian theory penetration into the STR that led to the sequence of Einstein's works on the generalization of relativity principle and to spreading the principle not only on inertial systems of reference, but on the various accelerated systems as well. Einstein applied the Principle of Equivalence to transform the knowledge not of classical mechanics only but the knowledge embodied in *both*, classical mechanics and STR. His theory of the static gravitational field, as well as his early attempts to generalize it, were nothing but a reinterpretation of the STR with the help of the introduction of accelerated frames of reference. His systematic treatment of such accelerated frames induced him to apply generalized Gaussian coordinates in order to describe the coordinate systems adapted to these frames of reference. It was then a natural step for him to consider the metric tensor. And with the introduction of the metric tensor Einstein constructed such a theoretical object that was capable of representing gravitational and inertial theoretical objects on the same footing.

By the beginning of 1912, Einstein realized that he would ultimately have to proceed beyond a scalar theory of gravitation. His strategy was to move carefully in a step-by-step manner towards a full dynamical theory. The first milestone of the research programme was to treat the "gravito-static" case, the gravitational analogue of electrostatics. However, he was already thinking about the second step, the "gravito-stationary" case, the gravitational analogue of magnetostatics. His ultimate aim was to advance a theory for time-dependent gravitational fields.

In March 1912 he confessed to Paul Ehrenfest:

> "The investigations of gravitational statics (point mechanics, electromagnetism, gravitostatics) are complete and satisfy me very much. I really believe that I have found a *part of the truth*. Now I am considering the dynamical case, again also proceeding from the more special to the more general case" (quoted from [20], 98; my italics).

As is well-known, in 1908-1911 Einstein had incessantly neglected research on gravitation, possibly because of his constant preoccupation with the problem of quanta. But this, however, seems to constitute only a part of true explanation. The remaining part refers to that he clearly realized how much work had to be done to arrive at an ultimate global theory. The latter should be able to embrace all the particular results obtained, the "parts of the truth" as Einstein called them, transforming them into the details of a great edifice. And, since Einstein himself was delved into the peculiarities of the theory of quanta, the problem of inventing the gravitation global theory scaffolds had fallen on Abraham and Nordström's broad shoulders.

Nevertheless, one has to keep in mind that even the pathways of creation of their ingenious theories were outlined by Einstein himself, starting with his ground-breaking 1907 paper. Indeed, one of the important STR consequences famously states that $E = mc^2$. Since, in a gravitational field, the energy of a particle depends on the value of the gravitational potential at the position of the particle, the equivalence of energy and mass suggests that:

1. either the particle's mass m;
2. or the speed of light c;
3. or both must also be a function of the gravitational potential.

All these possibilities, a dependence of the gravitational potential either on the speed of light c or on the inertial mass m, were later explored by Max Abraham ([2, 3, 14]) and Gunnar Nordström ([1, 33, 34]) respectively.

And first of all it became clear that one can easily invent such a Lorentz-invariant gravitation theory in which the inertial and gravitational masses are equal (Nordström, 1912-1914). Nordström's 1912 paper "*The principle of relativity and gravitation*" starts as follows:

> "Einstein's hypothesis that the speed of light c depends upon gravitational potential leads to considerable difficulties for the principle of relativity, as the discussion between Einstein and Abraham shows us. Hence, one is lead to ask if it would not be possible to replace Einstein's hypothesis with a different one, which leaves c constant and still adapts the theory of gravitation to the principle of relativity in such a way that gravitational and inertial mass are equal. I believe that I have found such a hypothesis, and I will present it in the following" (Nordström [1], 488).

Besides, initially Einstein's theory of static gravitational field did not offer even a hint at how the global theory should be contrived. On the contrary, a Göttingen theoretician, a true master of classical electrodynamics Max Abraham was one of the first scholars (along with Gustav Herglotz and Max Born) to propose that the four-dimensional line element, defining the infinitesimal distance between points in Minkowski space in terms of a constant metric, has to be replaced by a variable line element. Its functional dependence on the coordinates should be determined by a gravitational potential associated with the variable speed of light.

In a lecture presented in October 1912 and published the following year Abraham was one of the first (along with Henri Poincaré) to consider the possibility of gravitational waves in relativistic theories of gravitation. Furthermore, in 1912 G. Pavani had diligently calculated the perihelion shift of Mercury according to Abraham's theory, ascertaining a value that was approximately one third of the observed one. Abraham's theory thus made a more accurate prediction than even Einstein and Grossmann's '*Entwurf*'.

Einstein famously turned to the invention of the global gravitational theory only *after* the publication of Abraham's first vector gravitational theory. It should be punctuated that for static fields Abraham's theory coincides with Einstein's. But the most valuable result of the hybrid theories of Nordström and Abraham consisted in that the both theories contained ***extremely promising hints*** on how the global theory could be created [35].

At first, by letting the geometry of Minkowski space-time depend on the gravitational potential (Abraham). At second, by representing the gravitational potential not by a single function but by a ten-component queer theoretical object on a par with the stress-energy tensor and having this tensor as its source (Laue and Nordström). And, at third, by including an effect of the gravitational potential on the measurement of space and time (Nordström).

4. THE TENTATIVE CONTRIVANCE OF EINSTEIN AND GROSSMANN'S 'ENTWURF'

Consider Nordström's startling result produced with a help of M. Laue's achievements. The result draws on a fundamental problem of Maxwell's electrodynamics – the problem of electron's electromagnetic mass that owed so much to Abraham's 1909 achievements. If one computed total momentum and energy of the electromagnetic field of an electron, one arrives at the result universally accepted at that time:

$$\text{(Total field momentum)} = 4/3c^2 \text{ (Total field energy) (Electron velocity)}$$

Thus, as Poincaré and Einstein elucidated, there must also be stresses of a *non-electromagnetic* nature within the electron ('Poincaré's stresses'). So, the puzzle Max von Laue addressed in 1911 was to fix very general constraints under which the dynamic of such an electron would agree with the relativistic dynamics of point masses. While Hermann Minkowski had introduced the four-dimensional stress-energy tensor in the course of elaboration of four-dimensional methods in STR, his application of the tensor calculus was restricted to the special case of the electromagnetic field. Laue's writings concentrated on extending the application of this tensor to the most general domain [36 - 38]. The properties of this mathematical construct and its behavior under Lorentz transformations summarized a great deal of the then current knowledge of the behavior of stressed bodies.

As a result, Laue arrived at the expression for the stress-energy tensor $T_{\mu\nu}$ (μ,ν = 1,2,3,4) containing three basic blocs.

i. The first bloc represented the familiar three dimensional tensor p_{ik} (i,k = 1,2,3).
ii. The second bloc represented the momentum density **g** (g_x, g_y, g_z).
iii. The third bloc represented the energy flux **θ** (θ_x, θ_y, θz).

And the (T_{44}) component of the energy-momentum tensor represented energy. Einstein's Equivalence Principle prompted that each stress-tensor bloc should give *its own impact* into the gravitational field potentials, that is, each bloc is related to the gravitational potentials of its own. Hence there should be a multitude of gravitational potentials - scalar potentials, vector ones, *etc.* and not a single one .Thus the overall gravitational field potential should embrace several potentials and should in the most general case be described by a matrix, a tensor, since its components are transformed in the coordinate transformations like scalars, vectors, *etc*. The most pertinent analogy that played an important heuristic part was, of course, maxwellian electrodynamics with 4-dimensional electromagnetic field potential $A^{\mu} = (\mathbf{A}, \varphi)$. The latter, for static electromagnetic field, is reduced to static potential φ.

This hallmark of the genesis of GTR was later shrewdly indicated by such an eminent specialist in relativity as Sir Arthur Eddington who wrote:

> "Physical reality is the *synthesis* of all possible physical aspects of nature...Ultimately, if the 'synthesis' is performed correctly, physical quantities must be represented by *tensors* in 4-dimensional space-time" (quoted from [28], 198; my italics).

Note that in the Zurich Notebook, just before the publication of the '*Entwurf*', Einstein had freely worked with tensors. The traces of his work can be easily found in his unpublished review on the STR, probably prepared between 1912 and 1914 for "*Handbuch der Radiologie*". The heading of the section 3, dealing with vectors, tensors, *etc*., is: "*Some Concepts and Theorems of the Four-Dimensional Vector and Tensor Theories that Are Necessary for the Comprehension of Minkowski's Presentation of the Theory of Relativity*" (CPAE, vol.4, Doc.1). Moreover, one of the subsections is entitled "*The Stress-Energy Tensor of ElectromagneticProcesses*".

This hallmark was later diligently unfolded by an eminent Göttingen specialist in electrodynamics Max Abraham in his 1915 thought-provoking paper "*Recent Theories of Gravitation*". The paper contained such an important passage critically analyzing Einstein's and Grossmann's '*Entwurf*' that it is worth quoting in full.

> "The basic idea of the tensor theory of the gravitational field can be understood as follows. The energy density, which in a static field is determined by the divergence of the gradient of the gravitational potential, plays in the theory of relativity merely the role of one component of the resulting world tensor T; it is joined by nine other tensor components which characterize the energy flux and the stresses. The tensor theory assumes that, like the energy density (T_{44}), the remaining nine components $T_{\mu\nu}$ (μ, ν = 1,2,3,4) ***generate gravitational fields*** whose potentials $g_{\mu\nu}$ form a ten-tensor themselves

"(Abraham 1915 [14], 499; my italics).

The physical sense of the components is enunciated by Abraham below when he contends that the integration of Entwurf's field equations

> "is extraordinary difficult. Only the method of successive approximations promises success. In this one will usually take as a first approximation the solution that treats the field as static. Here, ***Einstein's theory becomes identical with Abraham's theory*** [...]"

In his Vienna lecture A. Einstein takes the normal values of the $g_{\mu\nu}$ as the first approximation: $g_{11} = g_{22} = g_{33} = 1$; $g_{44} = -c^2$, $g_{\mu\nu} = 0$ for $\mu \neq \nu$; he considers the deviations $g^*_{\mu\nu}$ from these normal values as quantities of first order, and arrives, by neglecting quantities of second order, at the following differential equations: $\Box g^*_{\mu\nu} = T^m_{\mu\nu}$. For incoherent motions of masses, the last (T^m_{44}) among the components of the material tensor T^m is the most important; it determines the potential $g^*_{44} = \Phi^g$. Then follow the components T^m_{14}, T^m_{24}, T^m_{34}, which are of first order in v/c; these determine the potentials g^*_{14}, g^*_{24}, g^*_{34}, which can be viewed as the components of a space vector – (1/c) U^g. The remaining components of T^m are of second order in v/c. If one neglects quantities of this order, then one only needs to consider those four potentials, and obtains for them the differential equations

$$\Box \Phi^g = c^2 \mu \qquad \textbf{(60a)}$$

$$\Box U^g = c^2 \mu \, (v/c) \qquad \textbf{(60b)}$$

where μ is the mass density.

Here the analogy with electrodynamics catches one's eye. Except for the sign, the field equations (60 a, b) agree with those that must be satisfied in the theory of electrons by the 'electromagnetic potentials', the scalar ond the vector othis approximation, *the Einstein-Grossmann tensor theory of the gravitational fields leads to the same results as the vector theory sketched in* (IA) [*i.e.* the theory of Abraham]" (Abraham 1915 [14], 500-501; my italics).

Abraham's startling conclusion can be re-affirmed by the consideration of Einstein's paper "*On the present state of the problem of gravitation*" (Einstein 1913c [39]). In this paper Einstein proceeded to show (p.1261) that the "*Entwurf*" theory reduced in suitable weak field approximation to a theory with a four-vector field potential that was formally analogous to maxwellian electrodynamics. It was this approximation that yielded the weak field effects now labeled as 'Machian'.

Just as Einstein himself had put it in the 8th paragraph "*The Newtonian Gravitational Field*", in the weak-field approximation "the Entwurf gravitational field" equations take the form

$$\Box g^*_{\mu\nu} = kT_{\mu\nu} \quad \textbf{(7c)}$$

where $\Box$ is the d'Alembertian wave operator.

"We obtain the Newtonian system insofar as we introduce the following approximations":

1. Only the mass flow is regarded as the field source.
2. The influence of the velocity of the field-generating masses is neglected, and hence the field is treated as static.
3. The velocity and acceleration components in the equations of motion of a material point are treated as small quantities, and only small quantities of the lowest order are retained. Finally, we also have to assume that at infinity the $g^*_{\mu\nu}$ vanish.

It then follows from (7c) and (8) that, if we write the Laplacian operator as Δ,

$\Delta g^*_{\mu\nu} = 0$ (unless $\mu = \nu = 4$)

$$\Delta g^*_{44} = kc^2 \rho_0 \quad \textbf{(7d)}$$

From this, as is well known, it follows that

$g^*_{\mu\nu} = 0$ (except in the case where $\mu = \nu = 4$)

$g^*_{\mu\nu} = kc^2/4\pi \int \rho_0 dv/r$", from which basic Newton's gravitation law is obtained:

$\ddot{x} = -(1/2)\, \partial g_{44}/\partial x$.

In the next approximation we "take into account the terms on the right-hand side of the system of equations (7c), which are proportional to the first power of the velocity of the field-producing masses. We then obtain the following instead of the system of equations (7d)"

$\Box g^*_{\mu\nu} = 0$ (if $\mu \neq 4$ and $\nu \neq 4$) [with $\Box$ representing the d'Alembertian wave operator]
$\Box g^*_{14} = -k\rho_0\, dx/dt$
$\Box g^*_{24} = -k\rho_0\, dy/dt$

$\Box g^*_{34} = -k\rho_0\, dz/dt$

$$\Box g^*_{44} = -\, kc^2\, \rho_0 \qquad \textbf{(7c)}$$

where $\Box$ is the d'Alembertian wave operator.

For a slowly moving point oOne can write these equations as follows in the usual three-dimensional vector notation:

$$\ddot{r} = -(1/2)\ \mathrm{grad}\ g_{44} + \dot{g} + [\,\dot{r}, o] \qquad \textbf{(1d)}$$

Here r = radius vector of the mass point

$\dot{r}$ = dr/dt *etc.*

g = a vector with the components g_{14}, g_{24}, g_{34},

o = curl ρ

If we denote the velocity of the field-producing masses (comp. $\dot{x}$, $\dot{y}$, $\dot{z}$) with v, then we can write (7c) in more concisely:

$\Box g = -k\rho_0\, v$

$$\Box g^*_{44} = kc^2\, \rho_0 \qquad \textbf{(7c')}$$

where $\Box$ is the d'Alembertian wave operator.

The equations (7c') and (1d) show how slowly moving masses influence each other according to the new gravitation theory. To a great extent, the equations correspond to those of electrodynamics, g_{44} corresponds to the scalar potential of electrical masses up to the sign [...]

G corresponds to the vector potential of electric currents; the second term of the right-hand side of (1d) , which corresponds to an electric field strength resulting from temporal change of the vector potential, yields precisely those induction effects, directed like acceleration, that we must expect based on our ideas regarding the inertia of energy. The vector o corresponds to the magnetic field strength (curl of the vector potential), so the last term in (1d) corresponds to the Lorentz force" (Einstein 1913c [39],564).

As his correspondence and the papers indicate, Einstein agreed with Nordström's assessment of the importance of Laue's work for gravitation theory. Furthermore, some pieces of his 1912 and 1913 papers (including his proposal to call T 'Laue's scalar') clearly indicate that he had personal contacts with Laue and discussed the stress-tensor problems with him. Such personal communication is compatible with the fact that both Einstein and Laue were teaching in Zurich, with Einstein at the ETH and Laue at the University of Zurich [5]. The same year Nordström also came to Zurich where supposedly he had communicated with the both scholars.

However what the connection between $T_{\mu\nu}$ and $g_{\mu\nu}$ should be? – An important hint is contained in Nordström's first 1912 paper: R= (k/2) T, where R is the fully contracted Riemann - Christoffel tensor and T is the trace of the stress-energy tensor. But if the equation is characteristic of scalar theory, the expression that generalizes it in a most natural and simple way should look like:

$k\ T_{\mu\nu} = \Gamma_{\mu\nu}$ (where $\Gamma_{\mu\nu}$ is a "contravariant second rank tensor formed by the derivatives of the fundamental tensor $g_{\mu\nu}$"). But these are exactly the gravitational field equations of the first metric theory – the theory of Einstein and Grossmann, published in 1913.

Just as Einstein himself have explained his strategy in "*Physical Foundations of a Theory of Gravitation*" –an address before the Swiss Society of Scientists on September 9, 1913, immediately following the '*Entwurf*',

> "the generalization of Poisson's equation (3) $\Delta\varphi = 4\pi k\rho$ is sought. On the other hand, the proportionality of energy and inertial mass that is obtained from the ordinary theory of relativity, and, on the other hand, the empirical proportionality of inertial and gravitational mass lead necessarily to the view that the same quantities that determine the energetic behavior of a system must also determine the gravitational effects of the system. From this we conclude that tensor $T_{\mu\nu}$ must appear in equations of gravitation we are seeking, in lieu of the density ρ of equation (3). We are therefore looking for equations that express the equality of two tensors, one of which is the given tensor $T_{\mu\nu}$, while the other comes from the fundamental tensor $g_{\mu\nu}$ through differential operations" (Einstein 1913b [31], 195).

It was immediately comprehended that in general the '*Entwurf*' equations are not covariant; they "remain covariant only with respect to linear orthogonal substitutions". However for a long time this peculiarity did not bother the authors. It indicates once more that the '*Enwurf*' field equations sprung out **not** from the Covariance Principle but represented a generalization of hybrid theories of Nordström and Abraham with a help of Laue's achievements.

Just as Einstein put it in his November 1913 letter to Paul Ehrenfest,

> "The gravitational affair has been clarified to my *complete satisfaction* (namely the circumstance that the equations of the gr. field are covariant only with respect to linear transformations). For it can be proved that generally covariant equations that determine the field completely from the matter tensor cannot exist at all. Can there be anything more beautiful than this, that the necessary specialization follows from the conservation laws" ([40], Doc. 481; my italics).

However, on the other hand, according to later reminiscences,

"The equivalence principle allows us...to introduce non-linear coordinate transformations in such a [4-dimensional] space [with pseudo-Euclidean metric]; that is, non-Cartesian ("curvilinear') coordinates. The pseudo-Euclidean metric then takes the general form:

> "$ds^2 = \sum g_{ik}\, dx_i dx_k$ summed over the indices i and k (from 1 to 4). These g_{ik} are then functions of the four coordinates, which according to the equivalence principle describe not only the metric but also the 'gravitational field'" (Einstein, quoted from [41], 55).

Thus, the question was raised on getting the mathematical apparatus dealing with such queer mathematical objects. From the mathematical point of view, the task was to discover a second order differential operator for the metric tensor covariant with respect to the largest possible class of coordinate transformations. In August 1912 Einstein left Prague, where he had taught for a year and a half, to become a full ETH professor. With Einstein's return to Zurich, he began a fruitful collaboration with his old friend Marcel Grossman. The collaboration ceased in 1914, when Einstein moved to Berlin to become a member of the Preussiche Akademie der Wissenschaften.

Grossmann's help was needed to treat the mathematical aspect of the problem. And he famously discovered that the exquisite mathematical apparatus was contrived at the end of the XIX-beginning of the XX-th century by Riemann, Levi - Civita, Ricci, Christoffel *et al.* That is why the first part of the '*Entwurf*' containing the gravitational field equations was written by A. Einstein, and only the second one – by M. Grossmann. Continuing the quotation from the 1955 reminiscences:

"I was made aware of these [works by Ricci, Levi- Civita *et al.*] by my friend Grossmann in Zurich, when I put to him the problem to investigate generally covariant tensors, whose components depend only on the derivative of the coefficients of the quadratic fundamental invariant.

He at once caught fire, although as a mathematician he had a somewhat skeptical stance towards physics... He went through the literature and soon discovered that the indicated mathematical problem had already been solved, in particular by Riemann, Ricci and Levi - Civita. This entire development was connected to the Gaussian theory of curved surfaces, in which for the first time systematic use was made of generalized coordinates" ([41], 15-16).

It is quite important for our study of the interrelations between Einstein's, Nordström's and Abraham's research programmes that the true commencement of the metric programme should be related to 1912, when in a paper published on 15 February 1912 as a reply to Einstein's critique, Abraham reconsidered the lines with which he had earlier considered the formalism of Hermann Minkowski. He introduced instead an infinitesimal line element with *variable metric*, thus actually extending Minkowski's space- time to a more general semi-Riemann manifold.

"In lines 16,17 of my note "On the Theory of Gravitation" an oversight has to be corrected which was brought to my attention by a friendly note from Herr A. Einstein. One should read there: 'we consider dx, dy, dz and du=icdt as components of a displacement $\overline{ds}$ in four-dimensional space'.

$$\text{Hence } ds^2 = dx^2 + dy^2 + dz^2 - c^2\, dt^2 \qquad \textbf{(12)}$$

is the square of the four-dimensional line element where the speed of light is determined by the equation (6) [*i.e.* $c^2/2 - c_0^2/2 = \Phi - \Phi_0$]" (Abraham 1912d [42], 1056).

In this way, Abraham was the first to introduce the mathematical representation of the gravitational potential that was to be at the heart of GTR, the general four-dimensional line element involving a variable metric tensor. Of course, for the time being, Abraham's startling expression remained as isolated mathematical contraption without deep physical meaning. Yet, on my view, it is this formula that was referred by Einstein in his 26 March letter to Michele Besso:

"At first (for 14 days!) I too was completely "bluffed" by the beauty

> and simplicity of his [Abraham's] formula" (CPAE [40], Doc. 377,436-437).

Driven possibly by Abraham's stubborn persistence, Einstein in May 1912 finally recognized that a generalized line element $ds^2 = dx^2 + dy^2 + dz^2 - c^2 dt^2$, as suggested by Abraham's shrewd comment of three months earlier, represents the key to a generally relativistic theory of gravitation.

Indeed, when Einstein had perfected the elaboration of his theory of a static gravitational field (Einstein 1912a [18],), he took notice of the fact that his expression for the motion in a static gravitational field takes an **exactly** the same form as that earlier given by Max Planck for a gravitation free Minkowski space-time, the only difference being that the speed of light c is now assumed to be variable:

$$\delta \int ds = \delta \{ \int \sqrt{(-dx^2 - dy^2 - dz^2 + c^2 dt^2)} \} = 0 \qquad \textbf{(1)}$$

where δ stands for the variation of the subsequent integral and c = c(x, y, z). Einstein was now quite enchanted by the possibility the equation (1) offered for a generalization beyond the special case of static gravitational fields, as was indicated by him in a famous "Note in proof" to his 1912a [18]:

> "This shows too – as has been shown for the ordinary theory of relativity by Planck – that the equations of analytical mechanics have a significant reading *far beyond* Newtonian mechanics. The Hamiltonian equation, which was the last one written down, gives an idea about how the equations of motion of the material point in a *dynamic* gravitational field are constructed" (Einstein 1912a [18], 120; my italics).

Now the pertinent question is what happens when one considers the motion of a point particle in the presence of *general, i.e. non-static* gravitational fields? – In that case the abovementioned Laue's results amended by the Equivalence Principle turned out to be of special importance. And in section 2 of the '*Entwurf*' Einstein takes the abovementioned variation principle as a starting point to argue that for non-static gravitational fields, too, one should expect equation (1) to describe the motions of point particles. Now the only difference is that the line element on the left-hand side of the equation has to be that defined by the metric tensor $g_{\mu\nu}$.

This was the first time the metric tensor was introduced in a published paper. Three months after the '*Entwurf*', Einstein submitted a paper to the 85th conference of the German Society for Scientists and Physicians. In this paper he explicitly maintained that "A free mass point moves in a straight and uniform line according to Eq. (1), where $ds^2 = \sum g_{ik}\, dx_i\, dx_k$". And finally on September 1913 in Vienna he presented a lecture elucidating the physical foundations of the '*Entwurf*' and those aforementioned conditions (1) - (4) which any relativistic theory of gravity should satisfy (Einstein 1913c [39]).

As a result, the main achievement of the second streak consisted in the invention of metric tensor; the latter appeared to be a hybrid (crossbred) theoretical object that unified two different research traditions – a physical tradition (scalar and vector theories of Einstein, Nordström and Abraham) and a mathematical one (geometrical results of Riemann, Christoffel, Levi-Civita *et al.*). As Einstein admitted in the letter to Lorentz,

"The gravitational field ($g_{\mu\nu}$) appears so to speak as the backbone on which everything hinges" (Einstein as quoted from [4], 19). Now the components of g_{ij} played a *dual* role: on the one hand, they represented the physical gravitational potentials and on the other they were coefficients in the expression of $ds^2 = \sum g_{ij}\, dx^i\, dx^j$. By means of contrivance of the crossbred object g_{ij} the interpenetration of geometry and physics commenced: physics became geometrized, and geometry was made empirical ([43], 267).

It should be stressed that the hallmark of Einstein's style that enabled him to generalize substantially different hybrid models of Abraham, Nordström and his own theory of static gravitational field was his *anti-dogmatism*. For instance, considering Einstein's efforts to create special relativity, Tetu Hirosige [44] shrewdly attributed Einstein's sensitivity to the inconsistencies between classical mechanics and Maxwellian electrodynamics to influence of Ernst Mach, whose writings supposedly helped the STR inventor to outdo dogmatic adherence to the mechanistic worldview. Einstein could therefore freely juxtapose Newtonian mechanics, statistical thermodynamics and Maxwellian electrodynamics without reducing one to the others [45].

Jürgen Renn and Robert Schulman [46] take Einstein's anti-dogmatism as a crucial hallmark of his scientific style of reasoning that enabled a young man to comprehend and even criticize the conceptual implications in the works of such luminaries as Lorentz, Hertz, Poincaré and Planck that they themselves were sometimes unable to discern. Unsurprisingly in their Proposal for Einstein's Membership in the Prussian Academy of Science (Berlin, 12 June 1913), M. Planck, W. Nernst, H. Rubens and E. Warburg famously maintained that

> "Apart from his great productivity, Einstein has a special talent for getting to the bottom of other scientists' newly emerging views and assertions, and for assessing their relationship to each other and to experience with surprising certainty"(Doc. N° 445 of Einstein [47], 338).

Likewise, as in the STR case, Einstein was able to freely juxtapose Nordström, Abraham's and his own non-metric theoretical schemes without reducing one to the others.

Note that *Einstein himself did not take the geometrization of the gravitational field as a major achievement of his research programme emphasizing that the GTR was no more and no less geometrical than Maxwell's theory of electromagnetism.* For instance, in his 8 April 1926 letter to Reichenbach, he famously confessed that

> "It is wrong to think that 'geometrization' is something essential. It is only a kind of crutch [Eselsbrücke] for the finding numerical results" (quoted from [48], 317).

Or, likewise, in Einstein's review of Emile Meyerson's book "*La deduction relativiste*", written in German by Einstein in 1927, it is contended that

> "The fact that the metric tensor is denoted as 'geometrical' is simply connected to the fact that this formal structure first appeared in the area of study denoted as 'geometry'. However, this is by no means a justification for denoting as 'geometry' every area of study in which this formal structure plays a role, not even for the sake of illustration one makes use of notions which one knows from geometry. Using a similar reasoning Maxwell and Hertz could have denoted the electromagnetic equations of the vacuum as 'geometrical' because the geometrical concept of a vector occurs in these equations" (quoted from [48], 318).

However, one more function of 'geometrization' should be taken into account. I think that it helped to construct a "*neutral theoretical language*" necessary to bring the substantially different theoretical traditions – that of Newtonian gravity and Special Relativity – under the same cover to compare and to reconcile them.

In this regard the following analogy with the genesis of Maxwellian electrodynamics seems to be especially appropriate.

In the course of electricity and magnetism unification, the first Maxwell's paper "*On Faraday's Lines of Force*" [1856 -1858] was dedicated to elaboration of the method of "analogies" (evoked by Kantian epistemology). The method decidedly rejects the "ontological" accounts looking for the "essences" of electrical and magnetic phenomena and asserting that 'in reality' electricity and magnetism are 'fields' and not 'action at a distance' phenomena, or *vice versa*. The crux of Maxwell's proposal was to comprehend Faraday's lines of force as a kind of *imaginary* tubes filled with *ideal* incompressible fluid.

> "I propose then, [...]; and lastly to show how by an extension of these methods, and the introduction of another idea due to Faraday, the laws of the attractions and inductive actions of magnets and currents may be clearly conceived, without making assumptions as to the *physical nature* of electricity, or adding anything to that which has been already proved by experiment.
>
> By referring everything to the purely geometrical idea of the motion of an imaginary fluid, I hope to attain generality and precision, and to avoid the dangers arising from a premature theory professing to explain the cause of the phenomena" (Maxwell [49], p.159; my italics).

It was important for a Kantian that the incompressible poison contrived has nothing to do with experimental reality. The constraints on the theory proposed consist in the demand that the mathematical constructs should not contradict each other. In all the other matters the method of physical analogies admits an unlimited freedom of imagination. Even the conservation laws could be freely broken down.

> "There is nothing self-contradictory in the conception of these sources where the fluid is created, and sinks where it is annihilated. The properties of the fluid are at our disposal, we have made it incompressible, and now we suppose it produced from nothing at certain points and reduced to nothing at others" (Maxwell [49], 162).

Maxwell punctuated the generality of the lines of force approach, for it could account for *any* kind of force. In particular, it does not exclude the force of action at a distance which varies inversely as the square of the distance, as force of gravity or as observed electric and magnetic phenomena.

And in the other parts of his path-breaking paper Maxwell elucidated the ways by which the idea of incompressible fluid motion could be applied to the sciences of statical electricity, permanent magnetism, magnetism of induction, and uniform galvanic currents. The core element of his splendid innovations consisted in constituting a language game with a "neutral language" for description and comparison of the consequences from the rival theories. Maxwell's "neutral language" was not Carnap's and Reichenbach's "observation language" springing out from the "protokolsatze" generalizations. Maxwell was clearly aware of the theory - laidenness of the observation data ("experimental laws already established, which have generally been expressed in the *language of other hypotheses*" (Maxwell [49], 162). He clearly understood that every observation necessarily carries the footprints of the theoretical language that helps to describe it. ("The daubing of untempered mortar", as he will call them later in his stupendous "*Helmholtz*" paper).

In order to compare and to unite in an ultimate theoretical scheme lacking contradictions all the results of the different experiments carrying the footprints of 'incommensurable' theoretical languages, it is necessary to construct an *artificial* and queer theoretical language equally distant from the languages of theories under comparison. In the case of electrodynamics this language turned out to be the solid state mechanics (with hydrodynamics as its part). Maxwell's ultimate aim was to rewrite all the known empirical and theoretical laws of electricity and magnetism using the neutral language and then to compare them in order to create a coherent system [50].

In a similar vein Einstein applied geometry as a kind of neutral theoretical language for the unification of gravitational and inertial phenomena. Moreover, the interpenetration of geometry and physics led to construction of the GTR fundamental theoretical scheme. The first stage of interpenetration resulted in the gravitational field equations of the '*Entwurf*': $R_{\mu\nu} = \varkappa T_{\mu\nu}$ with their ultimately simple premises of gravitational potentials being common partial derivatives of the metric tensor. Nevertheless, further penetration of physics into the geometry led to subtle modification of the plain theoretical scheme.

Finally, at the end of 1915 Einstein was able to admit to Arnold Sommerfeld:

> "You should not be mad at me that only today I reply to your friendly and interesting letter. The last month had been the most exciting and

> demanding time of my life; in any case also the most successful. I could not think of writing… After all faith in the results and methods of the earlier theory had dwindled, I clearly saw that only by taking up the general theory of covariants, that is, by taking up Riemann's covariant, a satisfying solution could be found… Of course it is easy to write down these generally covariant equations. But it is hard to see that they are the generalization of Poisson's equations and not easy to see that they allow satisfaction of the conservation laws… With Grossmann I had considered these equations three years earlier, up to the second term on the right hand side, but at the same time I came to the incorrect conclusion that they did not give the Newtonian approximation.
>
> The key to this solution was the realization that not $g^{ik} \frac{d}{dx} g_{ik}$ but the related Christoffel symbol $\{\frac{im}{l}\}$ should be taken as a natural expression for the 'components' of the gravitational field. Once one has come to this solution, then the above equation is the simplest conceivable, since one is not tempted to convert it by multiplying out the symbols for the sake of general interpretation.
>
> I experienced a glorious moment when not only Newton's theory in the first order approximation, but also the perihelion motion of Mercury (43" per century) in the second order approximation were retrieved" (Einstein's letter to Sommerfeld, 28 November 1915, quoted from [4], 30).

Yet let us turn to the details.

5. THE TRANSITION FROM 1913 ENTWURF TO THE ULTIMATE NOVEMBER 1915 THEORY OF GRAVITATION

The incessant interpenetration of geometry and physics eventually resulted in the construction of the GR fundamental theoretical scheme. The first stage of interpenetration resulted in the gravitational field equations of the '*Entwurf*': $R_{\mu\nu} = \varkappa T_{\mu\nu}$ with their ultimately simple premises of gravitational potentials being common partial derivatives of metric. However the further interplay of physics and geometry led to skillful modification of the plain scheme. In a series of four publications from November 1915, submitted to the Prussian Academy of Science on the 4th, 11th, 18th and 25th, Einstein gradually replaced the '*Entwurf*' by a full-blooded metric theory of gravitation, solving incidentally the problem of

Mercury's perihelion. To comprehend Einstein's reasoning in passing from the '*Entwurf*' to the GR one has to address the abovementioned Einstein's synthetic dual strategy that embraced and successfully reconciled both a physical and mathematical approaches to finding the ultimate field equations. The '***dual method***' suggested that Einstein "inductively" proceeded first and foremost from a set of the following constraints of a *physical* nature:

> (P1) *The Newtonian limit.* Any relativistic theory of gravitation must obviously correspond to Newtonian theory in the case of weak and static gravitational fields. So, in this limit one must recover a Poisson-like equation for a scalar potential from the field equations.
> (P2) *Conservation of energy and momenta.* On the other hand, the complementary 'top-down' (van Dongen) or 'deductive' *mathematical* approach originated from the following two requirements.
> (M1) *The Principle of Equivalence.* Observations made in a uniformly accelerated system are equivalent (in the abovementioned sense) to observations made in a homogeneous gravitational field.
> (M2) *Generalization of the Principle of Relativity.* The laws of physics are identical for relatively inertially moving observers to observers in accelerated motion. The (M1 +M2) jointly imply that the field equations have to be generally covariant. Thus, a natural starting point for the mathematical approach would be to scrutinize a generally covariant mathematical abstract object that was known from the mathematics literature. The key point is that the 'dual strategy' necessarily employed *both* physical and mathematical approaches: "it was an *iterative* process that began with trying out one approach, and then checking results with the demands or results of the other approach" ([4], 11).

Yet the physical approach first prevailed and leads to the '*Entwurf*'; but then the mathematics approach took over and incessantly led to 1915 November General Relativity. It is interesting to note that in the "*Zurich Notebook*" Einstein already had the gravitational field equations that he would publish in November 1915. He had been necessarily led to them by the abovementioned 'interplay' of the dual method; however, they appeared to be inconsistent with a coherent system of 'physical' constrains. First and foremost, they did not provide the Newton metric $g_{\mu\nu}$ = diag (g_{00}, -1.-1,-1) in the weak-field approximation. On the contrary, an obvious advantage of the '*Entwurf*'' equations was that, when taking the Newtonian limit, one was not obliged to insert the notorious 'harmonic coordinates' restrictions to recover the Poisson equations in its familiar to Einstein form, and blatant inconsistencies with the Newton metric $g_{\mu\nu}$ = diag (g_{00}, -1.-1,-1) were therefore not to be expected. Moreover, Einstein's confidence in the '*Entwurf*' was strengthened by a notorious 'hole argument' ("happenings in the gravitational field cannot be uniquely determined by generally covariant differential equations for the gravitational field").

So, eventually in November 1913 Einstein could confess to his former assistant Ludwig Hopf:

> "I am now very satisfied with the gravitation theory. The fact that the gravitational equations are not generally covariant, something that quite disturbed me for a while, is unavoidable; it can easily be shown that a theory with generally covariant field equations cannot exist if the demand is made that the field is mathematically completely determined by matter" (2 November 1913, A. Einstein to L. Hopf; quoted from [4], 22).

Yet in the light of the dual strategy and the synthetic character that a true relativistic theory of gravity should possess it is no wonder that the absence of generally covariant field equations was strongly criticized by Max Abraham and Gustav Mie. However, initially Einstein was blind to such criticism.

Nevertheless, A. Einstein and M. Besso applied the metric $g_{\mu\nu}$ = diag (g_{00}, -1.-1,-1) to calculate for the first time the anomalous perihelion shift of Mercury. The result appeared to be a factor of 2, 4 off from the observed value. Much distressed, Einstein would soon give up on the troubled theory; yet eventually his reasons were threefold. As he later wrote to Arnold Sommerfeld,

> "[At first] I proved that the gravitational field on a uniformly rotating system does not satisfy the gravitational field equations.
>
> [At second] the motion of the perihelion of Mercury came out as 18'' instead of 45'' per century.
>
> [At third] the covariance argument in my paper of last year does not give the Hamiltonian function H. When suitably generalized it allows an arbitrary H" (Einstein to A. Sommerfeld, 28 November 1915, Doc.153; quoted from [4], 26).

As a result, Einstein found a way out of the '*Entwurf*' predicament: he returned to the half-forgotten mathematical strategy and to the requirement of general covariance that he harshly abandoned in the '*Zurich Notebook*'. Subsequently, in the week after 11 November 1915, Einstein redid the calculation of Mercury's perihelion shift. He realized that covariant equations $R_{\mu\nu}$ - (R/2) $g_{\mu\nu}$ = $kT_{\mu\nu}$ cannot be consistent with the Newton metric $g_{\mu\nu}$ = diag (g_{00}, -1.-1,-1). Now he was forced to recognize the substantially tensor character of the theory and had to allow the spatial components of the metric tensor to be functions of the coordinates too: $g_{\mu\nu}$ = diag {$g_{00}(x)$, $g_{11}(x)$, $g_{22}(x)$, $g_{33}(x)$}. The ds^2 equation that Einstein used in the November 1915 paper to calculate Mercury's motion strongly resembled the Schwarzchild black hole metric with it's g_{00} = (1-2m/r) and the g_{11}-g_{33} terms depending on r and m. Digging still deeper into the mathematical background, Einstein elicited, at last, the correct value of the perihelion shift: he famously found 43'', which confirmed his desired 45'' ± 5'' per century.

Thus, it appeared to be the reversion to the mathematical requirement of general covariance that finally brought salvation and that eventually relieved Einstein of his prejudices regarding the Newtonian limit $g_{\mu\nu}$ = diag (g_{00}, -1.-1,-1).He eloquently expressed his amazement at this 'enormous achievement' in a famous letter to Michele Besso:

> "Read the articles! They bring the final salvation to misery. The most joyful aspect is *the accordance between the perihelion motion and general covariance*, the most striking however the circumstance that Newton's theory for the field is already incorrect at the first order of the equation (the terms g_{11}-g_{33} arise). The simplicity of Newton's theory is only due to the fact that the g_{11}-g_{33} do not arise in the first approximation to the equations of motion for a point mass" (21 December 1915, Einstein to M. Besso, quoted from [4], 29; my italics).

In the light of the proposed reconstruction of the GR genesis the abovementioned "accordance between the perihelion motion and general covariance" appears to be obvious. Einstein could arrive at correct explanation of the abnormal perihelion motion (grounded on the understanding that there are many different *gravitational* forces that act not only along the plane of Mercury rotation round the Sun) only when he had fully realized the genuinely tensor character of the gravitational field. That is, when he realized that the $g_{\mu\nu}$ is a host not only of scalar components like g_{00}(r, m), but of the g_{11}, g_{22}, g_{33} components as well, each of which also depend on r and m. It is important to acknowledge the following. The ultimate GR equations were contrived only in November 1915, so the whole two years had passed after the blatant discrepancy between the '*Entwurf*' and the perihelion motion was found in 1913. And though Einstein indeed mentioned this discrepancy as one of his basic motivations for the 'Entwurf-GR transition' (see the letter to Arnold Sommerfeld, 28 November 1915, quoted in [4], 31), the things were not so simple. The '*Entwurf*' really could not explain *all* the abnormal motion, but Einstein quite reasonably did not consider this fact as an ultimate refutation of his theory since the remaining unexplained parts of the precession could be due to other, non-gravitational factors. Einstein and Besso had found the faulty value already in June 1913, but during the two years – up to September 1915 – Einstein constantly and persistently expressed confidence in the '*Entwurf*' theory. (See, for instance, his letter to Michele Besso from 10 March 1914 [4], 23). On my view, the main thrust in passing from the '*Entwurf*' to the GR came from his dual strategy inducing him to *reconcile* physical and mathematical approaches. And only *after* he efficiently reconciled them, that is, only after he realized the general covariance of the gravitational field equations, could he turn to correct explanation of the Mercury perihelion. Hence the decisive impetus in constructing the General Relativity came from the attempts to reconcile physical and mathematical approaches, embodied in the Abraham, Nordstrom and Einstein's crossbred theoretical models. Trustworthy explanation of Mercury perihelion motion appeared to be a by-product of strenuous reconciliation efforts.

6. CONCLUSIONS

To recapitulate, the reconstruction proposed enables to highlight the following discernible hallmarks of the process of the GTR genesis that are usually obfuscated by the other studies and to arrive at a more comprehensive account of the *intertheoretic context* of Einsteinian revolution. (One should always keep in mind that real creative science, and namely Einstein, Poincaré, Lorentz, Abraham's *et al. modus operandi* is always messier and more complicated than philosophers of science like to fancy or scientists themselves like to assert in numerous festschrifts, commemorative volumes, *etc.*).

i. Invention of relativistic theory of gravity had dawned with the crossbred object construction in Einstein's 1907 paper. The crossbred object introduction – the introduction of inertial and simultaneously gravitational mass ($E = mc^2$) – led *perforce* to STR methods penetration into Newtonian theory of gravity and to reverse penetration of Newtonian gravity methods into the STR. As a result, the both theories were radically restructured and the corresponding changes in both of them were set up. The changes were epitomized in the peculiar sequences of crossbred models, the byproducts of the transformation performed.
   a. On the one hand, an inevitable consequence of the STR invasion into Newtonian theory of gravity turned out to be Nordström's and Abraham's scientific research programmes.

b. On the other hand, no less inevitable, owing to the Equivalence Principle, was the Newtonian theory invasion into the STR that led to the sequence of Einstein's papers on the generalization of relativity principle and to spreading the principle not only on inertial systems of reference, but on the various accelerated systems as well. But the most valuable result of the hybrid theories of Nordström and Abraham consisted in that the both theories contained *very promising hints* on how the global theory could be created. Hence the climax of the stage was Einstein's proposal and apprehension of the Equivalence Principle that became one of firm GTR heuristic foundations.

ii. The basic GTR ideal model was constructed due to the unification of the hybrid models of Einstein, Nordström and Abraham invented within *different* research programmes. The pivotal hallmark of Einstein's style of reasoning was his open-mindedness, his consistent anti-dogmatism that made possible for Einstein to freely juxtapose Nordström, Abraham's and his own non-metric theoretical schemes without reducing one to other.

iii. It is this hallmark that makes it possible to comprehend the true reasons for Einstein's victory over the rival programmes of Nordström and Abraham. The '*Entwurf*' and the GTR 'incorporated' Nordström and Abraham's theories in a higher level theory. Thus, the true reason for Einstein's victory over the rival programmes does not consist in that he resolutely rejected the alternative theories. On the contrary, it consists in that he subsumed them all into a new framework in substantially modified forms.

iv. The monumental volumes (Renn *et al.*, 2007), to which I repeatedly referred, are permeated with the idea that GTR "was not just a one-man affair and a lucky individual discovery but the result of a *profound transformation* of the extended system of knowledge of classical physics". According to the authors, the genesis of the GTR can be taken as an example of a 'Copernican process' that is repeatedly observable at decisive moments in the history of science.

> "This process implies that marginal elements of a complex architecture of Knowledge torn asunder by inner tensions become the starting points for a reconstruction as a result of which a new structure emerges that is, however, essentially composed of the building blocks already available. This process may be designated as a "Copernican process". Indeed conceptual turnovers unfold in a similar way to the revolution of Copernicus, who created a new world system by placing an initially marginal celestial body – the sun – into the center , while, instead of starting with a *tabula rasa*, retaining the previously elaborated complex machinery of planetary astronomy" ([6], 47).

In a sense this chapter can be taken as an attempt to *explicate* and *furnish* the concept of the aforementioned 'Copernicus process', providing a *mechanism* of the classical paradigms simultaneous profound transformation. The hallmark of the mechanism is the construction of the crossbred systems from the basic theoretical objects of 'old' mature theories [21]. These crossbred systems are subsequently generalized to coin a structure of a new mature theory. Any mature physical theory can be empirically successful if it represents a kind of an 'entwurf' that can be explicated by the results of new experiments that increase its empirical content. In this way the GTR could become such a vehicle for production of new gravitational propositions since it managed to unify and intertwine the 'old' classical traditions of Newtonian mechanics (including Newtonian gravity) and the STR in such a way that they ceased to contradict each other and their newly moulded fusion could successfully explain and predict new experimental data.

REFERENCES

[1] G. Nordström, "Relativitätsprinzip und Gravitation", *Phys. Z.,* vol. 13, pp. 1126-1129, 2007.

[2] M. Abraham, "Das Elementargesetz der Gravitation", *Phys. Z.,* vol. 13, pp. 4-5, 1912.

[3] M. Abraham, "Zur Theorie der Gravitation", *Phys. Z.,* vol. 13, pp. 1-4, 1912.

[4] J. Van Dongen, *Einstein's Unification.* Cambridge University Press: Cambridge, 2010. [http://dx.doi.org/10.1017/CBO9780511781377]

[5] J.D. Norton, "Einstein, Nordström and the early demise of scalar, Lorentz covariant theories of gravitation", In: *The Genesis of General Relativity. Vol. 3. Gravitation in the Twilight of Classical Physics: Between Mechanics, Field Theory and Astronomy.* Dordrecht: Springer, pp. 413-488, 2007.

[6] J. Renn, and T. Sauer, "Pathways out of Classical Physics: Einstein's Double Strategy in his Search for the Gravitational Field Equations." *In: The Genesis of General Relativity.* vol. 1. Springer: Dordrecht, 2007, pp. 113-312.

[7] J. Renn, "The summit almost scaled: Max Abraham as a pioneer of a relativistic theories of gravitation." *In: The Genesis of General Relativity. Gravitation in the Twilight of Classical Physics: Between Mechanics, Field Theory and Astronomy.* vol. 3. Springer: Dordrecht, 2007, pp. 305-330.

[8] A. Einstein, and M. Grossmann, "Entwurf einer verallgemeinerten Relativitätstheorie und Theorie der Gravitation", *Zeitschrift fur Mathematik und Physik,* vol. 62, pp. 225-261, 1913.

[9] A. Einstein, and A. Fokker, "Die Nordströmsche Gravitationtheorie vom Standpunkt des absoluten Differentialkalkulis", *Ann. Phys.,* vol. 44, pp. 321-328, 1914. [http://dx.doi.org/10.1002/andp.19143491009]

[10] A. Einstein, "Die Feldgleichungen der Gravitation", Preussiche Akademie der Wissenschaften", *Sitzungberichte,* vol. 48, pp. 844-847, 1915.

[11] W. Pauli, *Theory of Relativity.* Dover Publications: N.Y., 1981.

[12] L.D. Landau, and E.M. Lifshitz, *Theory of Fields. L.* Pergamon Press, 1983.

[13] A. Einstein, "Näherungsweise Integration der Feldgleichungen der Gravitation"", *Preussiche Akademie der Wissenschaften Sitzungsberichte,* vol. 1, pp. 688-696, 1916.

[14] M. Abraham, "Neuere Gravitationstheories", *Jahrbuch der Radioaktivität und Elektronik,* vol. 11, pp. 470-520, 1915.

[15] J. Stachel, "The Rigidly Rotating Disk as the 'Missing Link' in the History of General Relativity" *In: Einstein and the History of General Relativity. (Einstein Studies, Vol. 1)* Birkhauser: Boston, Basel and Berlin, 1989, pp. 48-62.

[16] Ja. B. Zel'dovich, and I.D. Novikov, *Stars and Relativity.* Pergamon Press: London, 1973.

[17] M. Abraham, "Der Freie Fall", *Phys. Z.,* vol. 13, pp. 310-311, 1912.

[18] A. Einstein, "Zur Theorie des statischen Gravitationfeldes", *Ann. Phys.,* vol. 38, pp. 443-458, 1912. [http://dx.doi.org/10.1002/andp.19123430709]

[19] M. Janssen, J.D. Norton, and J. Renn, "Introduction to volumes 1 and 2: the Zurich Notebook", In: *Boston Studies in the Philosophy of Science.* vol. 250. Springer: Dordrecht, 2007, pp. 7-20.

[20] J. Renn, "Classical Physics in Disarray. The Emergence of the Riddle of Gravitation." *In: The Genesis of General Relativity.* vol. 1-2. Springer: Dordrecht, 2007, pp. 21-84.

[21] R.M. Nugayev, *Reconstruction of Mature Theory Change: A Theory-Change Model.* Peter Lang: Frankfurt am Main, 1999.

[22] V.S. Stepin, *Theoretical Knowledge. Dordrecht-Reidel.* Springer, 2005. [http://dx.doi.org/10.1007/1-4020-3046-0]

[23] L. Corry, *David Hilbert and the Axiomatization of Physics (1898-1918).* Kluwer Academic Publishers: Dordrecht, 2004. [http://dx.doi.org/10.1007/978-1-4020-2778-9]

[24] V. Petkov, "Physics as Space-time Geometry, Chapter 8". *In: Handbook of Space-time* Springer: Heidelberg, 2014, pp. 141-163.

[25] A. Einstein, "Über das Relativitätsprinzip und die aus demselben gezogenen Folgerungen", *Jachrbuch der Radioaktivität und Elektronik*, vol. 4, pp. 411-462, December 1907. Translated by Anna Beck in: *The Collected papers of Albert Einstein, vol. 2, The Swiss Years: Writings, 1900-1909*, pp. 252-311.

[26] A. Pais, *Subtle is the Lord. The Science and Life of Albert Einstein.* Oxford University Press: Oxford, 1982.

[27] A. Papapetrou, "Spinning Test Particle in General Relativity", *Proc. Roy. Soc. A,* vol. A 209, no. 1097, pp. 248-258, 1951. [http://dx.doi.org/10.1098/rspa.1951.0200]

[28] T. Ryckman, *The Reign of Relativity: Philosophy in Physics 1915-1925.* Oxford University Press: Oxford, 2005. [http://dx.doi.org/10.1093/0195177177.001.0001]

[29] M. Janssen, "The twins and the bucket: How Einstein made gravity rather than motion relative in general relativity", *Stud. Hist. Philos. Sci. Part Stud. Hist. Philos. Mod. Phys.,* vol. 43, pp. 159-175, 2012. [http://dx.doi.org/10.1016/j.shpsb.2012.01.003]

[30] J.D. Norton, "What was Einstein's Principle of Equivalence Principle?" *In: Einstein's Studies.* vol. 1. Birkhauser: Boston, 1989, pp. 5-47.

[31] A. Einstein, "Physikalische Grundlagen einer Gravitationtheori", *Naturforchende Gesellchaft Vierteljahrsschrift Zurich,* vol. 58, pp. 284-290, 1913.

[32] A. Einstein, "Relativität und Gravitation. Erwiderung auf eine Bemerkung von M. Abraham", *Ann. Phys.,* vol. 38, pp. 1059-1064, 1912. [http://dx.doi.org/10.1002/andp.19123431014]

[33] G. Nordström, "Trage und Schwere Masse in der Relativitätsmechanik", *Ann. Phys.,* vol. 40, pp. 856-878, 1913. [http://dx.doi.org/10.1002/andp.19133450503]

[34] G. Nordström, "Zur Theorie der Gravitation vom Standpukt des Relativitätsprinzip", *Ann. Phys.,* vol. 42, pp. 533-534, 1913. [http://dx.doi.org/10.1002/andp.19133471303]

[35] M. Abraham, "Zur elektromagnetischen Mechanik", *Phys. Z.,* vol. 10, pp. 737-741, 1909.

[36] M. Laue, "Zur Dynamik der Relativitätstheorie", *Ann. Phys.,* vol. 35, pp. 524-542, 1911. [http://dx.doi.org/10.1002/andp.19113400808]

[37] M. Laue, *Das Relativitätsprinzip*. Braunschweig: Friedrich Vieweg und Sohn, 1911.

[38] M. Laue, "Ein Beispiel zur Dynamik der Relativitätstheorie", *Verhandlungen der deutsches Physikalische Gesellchaft*, vol. 13, pp. 513-518, January-December 1911

[39] A. Einstein, "Zum gegenwartigen Stande des Gravitationsproblems", *Phys. Z.*, vol. 14, pp. 1249-1262, September 1913. Translated in: *The Genesis of General Relativity. Gravitation in the Twilight of Classical Physics: Between Mechanics, Field Theory and Astronomy*. Vol. 3. Dordrecht: Springer, 2007.

[40] M.J. Klein, and A. J. Kox, *The Collected Papers of Albert Einstein. Vol. 5. The Swiss Years: Correspondence, 1902-1914*. Princeton, NJ: Princeton University Press, 1993.

[41] C. Seelig, *Helle Zeiten – Dunkle Zeiten: In memoriam Albert Einstein.* Europa Verlag: Zurich, Stuttgart, Wien, 1957.

[42] M. Abraham, "Nochmals Relativität und Gravitation. Bemerkung zu A. Einsteins Erwiderung", *Ann. Phys.,* vol. 38, pp. 1056-1058, 1912. [http://dx.doi.org/10.1002/andp.19123431013]

[43] E. Zahar, *Einstein's Revolution: A Study in Heuristic*. La Salle: Open Court, 1989.

[44] T. Hirosige, "The ether problem, the mechanistic world view and the origins of the theory of relativity", *Hist. Stud. Phys. Sci.,* vol. 7, pp. 3-82, 1976. [http://dx.doi.org/10.2307/27757354]

[45] R.M. Nugayev, "The History of Quantum Theory as a Decisive Argument Favoring Einstein over Lorentz", *Philos. Sci.,* vol. 52, pp. 44-63, 1985. [http://dx.doi.org/10.1086/289221]

[46] J. Renn, and R. Schulmann, "Introduction." *In: Albert Einstein/ Mileva Maric. The Love Letters.* Princeton University Press: Princeton, NJ, 1992, pp. 3-22.

[47] *The Early Years, 1879-1902.* vol. 1. Princeton University Press: Princeton, NJ, 1987.J. Stachel, Ed., The Collected Papers of Albert Einstein.

[48] D. Lehmkuhl, "Why Einstein did not believe that general relativity geometrizes gravity", *Stud. Hist. Philos. M.P.,* vol. 46, pp. 316-326, 2014. [http://dx.doi.org/10.1016/j.shpsb.2013.08.002]

[49] J.C. Maxwell, "On Faraday's Lines of Force". *Transactions of the Cambridge Philosophical Society*, vol. X, part 1, read on December 10, 1855 and on February 11, 1856 and published in 1858. Reprinted in: *The Scientific Papers of James Clerk Maxwell*, 1890, vol. 1, Cambridge. University Press and New York: Dover, 1952, pp. 155-229

[50] R.M. Nugayev, "Communicative Rationality of the Maxwellian Revolution", *Found. Sci.,* vol. 20, no. 4, pp. 447-478, 2015. [http://dx.doi.org/10.1007/s10699-015-9417-z]

SUMMARY

It is argued that to overcome the shortcomings of convergent realism one has to turn from classical (convergent) or 'metaphysical' realism to non-classical or 'internal' one, grounded on neo-Kantian epistemology, and to coherent theory of truth. But then she has to grapple with a problem: where does the objectivity of science comes from? It is contended that to propose a sober answer one has to revise the common concept of growth of knowledge on the basis of the epistemological model of mature theory change that hinges upon 'old' theories encounter and interaction. According to the lucid model, each paradigm is a local viewpoint determined by the peculiarities of a culture into which its creator was submerged. However, an encounter of different paradigms leads to their interaction; as a result, the crossbred theoretical objects are constructed. Through these systems the infiltration of one paradigm on the other's domain takes place. After the old paradigms' grinding the new ones emerge that reconcile to each other much better than the old ones.

It should be stressed, that according to the epistemological model, *radical breakthroughs in science were not due to invention of new paradigms or the creation of new ideas ex nihilo, but rather to the long-term processes of the reconciliation, interpenetration and intertwinement of 'old' research traditions preceding such breaks.*

To elucidate the growth of knowledge model proposed, the inter-theoretical context of the Copernican revolution is scrutinized. It is contended that the Copernican revolution can be taken as bringing to light the dualism between mathematical astronomy and Aristotelian qualitative physics and the corresponding gradual efforts to eliminate it. The works of Copernicus, Galileo, Kepler and Newton were all the stages of the mathematics descendance from skies to earth and reciprocal extrapolation of earth physics on divine phenomena.

The second significant case study is the Maxwellian revolution. The arguments are strengthened in favor of the tenet according to which the genesis of maxwellian electrodynamics can be comprehended as a result of the old pre-maxwellian research traditions encounter, intertwinement and reconciliation. The research programmes to coordinate with each other had been the electrodynamics of Ampére-Weber, the wave theory of Young-Fresnél and Faraday's programme. It is maintained that Maxwell's unification design could be successfully implemented since his programme had assimilated the ideas of the Ampére-Weber programme, as well as the presuppositions of the programmes of Young-Fresnél and Faraday. Eventually Maxwell's victory over his outstanding rivals became possible since the core of Maxwell's unification strategy had been permeated by the spirit of Kantian epistemology. Maxwell had put forward as a basic synthetic principle the tenet that radically differed from that of the rival approaches by its open, flexible and contra-ontological, genuinely Kantian character.

The third example thoroughly analyzed is Einstein's 1905 'annus mirabilis'(wonder year). It is maintained that to figure out in what way different parts of Einstein's 1905 'annus mirabilis' writings hang together one has to take into account Einstein's strive for unity evinced in his persistent efforts to reconcile maxwellian electrodynamics, statistical mechanics and thermodynamics. It is exposed that special relativity turns out to be a mere milestone of implementation of maxwellian electrodynamics, statistical mechanics and thermodynamics unification programme. Since the luminiferous ether conception was a snag for Einstein's statistical thermodynamics design, the leading part in the programme was played by Einstein's 1905 light quanta hypothesis. Though influenced in critical stand against the classical physics research tradition by David Hume and Ernst Mach, Einstein hold an overall eclectic philosophical stance. Yet it is exhibited that when it related to creative momenta, Einstein's actual 1905 standpoint was close to neo-Kantian epistemology. It is contended that the most important Kantian concept apt to conceive Einstein's special theory of relativity creation and all his 1905 papers *as a whole*, as well as the *order* of their arrangement is Kant's regulative idea of the systematic Unity of Nature.

And, at last, the inter-theoretical aspect of the General Theory of Relativity genesis and advancement as an instance of the proposed epistemological model is comprehended. I strengthen arguments in favor of the tenet that the dynamic creation of General Theory of Relativity had been continually governed by internal tensions between two research traditions, that of Einstein's special theory of relativity and Newton's gravity. The encounter of the traditions, their interpenetration and intertwinement entailed coming into being the rival programmes of relativistic theory of gravity construction: of the programmes of Abraham and Nordström – on the one hand – and Einstein's programme – on the other. The encounter provided the construction of the hybrid domain at first with an irregular set of theoretical models. The rival approaches of Einstein, Abraham and Nordstrom grew closer and eventually were turned out by Einstein into complementary 'mathematical' and 'physical' strategies of a systematic synthetic research programme that engendered the '*Entwurf*' and the General Relativity. Step by step, on eliminating the contradictions between the models contrived, the hybrid set was put into order by Einstein *via* the Principle of Equivalence. It is contended that the main reason for the General Relativity 'victory' over the rival programmes of Abraham and Nordström was a synthetic character of Einstein's programme. Einstein's programme did supersede the rival ones because it did ingeniously assimilate some ideas of the Nordström programme as well as some viable premises of the programme of Abraham. Einstein subsumed them all into a new framework. Hence the decisive impetus in constructing the General Relativity came from the attempts to reconcile physical and mathematical approaches, embodied in the Abraham, Nordstrom and Einstein's crossbred theoretical models. Trustworthy explanation of Mercury perihelion motion appeared to be a by-product of strenuous reconciliation efforts.

SUBJECT INDEX

A

B

C

Rinat Nugayev
All rights reserved-© 2018 Bentham Science Publishers

D

E

F

G

H

I

J

K

L

M

N

O

P

S

V

W

Y

Z

www.ingramcontent.com/pod-product-compliance
Lightning Source LLC
LaVergne TN
LVHW070951210926
848482LV00002B/174
9781681086361